# THE
# EXCIPLEX

Proceedings of a meeting held under the auspices of the
Photochemistry Unit, University of Western Ontario,
May 28–31, 1974

*Organizing committee:*
D. R. Arnold
M. Gordon
P. de Mayo
W. R. Ware

# THE
# EXCIPLEX

edited by

## M. Gordon
*Merck Sharp & Dohme Canada Ltd.-Isotopes*
*Pointe Claire-Dorval, Quebec, Canada*

## W.R. Ware
*Photochemistry Unit*
*University of Western Ontario*
*London, Ontario, Canada*

Academic Press Inc., New York   San Francisco   London   1975

A Subsidiary of Harcourt Brace Jovanovich, Publishers

ACADEMIC PRESS, INC.
111 Fifth Avenue, New York, New York 10003

*United Kingdom Edition published by*
ACADEMIC PRESS, INC. (LONDON) LTD.
24/28 Oval Road, London NW1

Library of Congress Cataloging in Publication Data

International Exciplex Conference, University of
        Western Ontario, 1974.
    The exciplex.

    "Papers presented at the International Exciplex
Conference held at the University of Western Ontario,
May 28-31, 1974."
        Bibliography: p.
        Includes index.
        1.    Excited state chemistry—Congresses.    2.    Photo-
chemistry—Congresses.        I.    Gordon, M.        II.    Title.
QD461.5.I57 1974              541'.35              74-27524
ISBN 0-12-290650-0

# CONTENTS

CONTENTS

# CONTRIBUTORS

A.J. BARD—Department of Chemistry, The University of Texas at Austin, Austin, Texas 78712

J.B. BIRKS—The Schuster Laboratory, University of Manchester, Manchester, United Kingdom

F.A. CARROLL—Department of Chemistry, University of North Carolina, Chapel Hill, North Carolina 27514

E.A. CHANDROSS—Bell Laboratories, Murray Hill, New Jersey 07974

D.A.ERSFELD—Chemistry Department, Michigan State University, East Lansing, Michigan 48824

T.R. EVANS—Research Laboratories, Eastman Kodak Company, Rochester, New York 14650

S. FARID—Research Laboratories, Eastman Kodak Company, Rochester, New York 14650

T H. FÖRSTER—late of the Institut für Physikalische Chemie der Universität Stuttgart, Germany

S.E. HARTMAN—Research Laboratories, Eastman Kodak Company, Rochester, New York 14650

O. JAENICKE—Research Laboratories, Eastman Kodak Company, Rochester, New York 17650

A. LEDWITH—Donnan Laboratories, University of Liverpool, Liverpool, L69 3BX, England

N. MATAGA—Department of Chemistry, Faculty of Engineering Science, Osaka University, Osaka, Japan

*Present Address: Merck Sharp and Dohme Canada Ltd., Isotopes Pointe Claire-Dorval, Quebec, Canada

J. MICHL–Department of Chemistry, University of Utah, Salt Lake City, Utah 84112

N. ORBACH–Department of Physical Chemistry, The Hebrew University, Jerusalem, Israel

M. OTTOLENGHI–Department of Physical Chemistry, The Hebrew University, Jerusalem, Israel

S.M. PARK–Department of Chemistry, The University of Texas at Austin, Austin, Texas 78712

R.D. POSHUSTA–Chemical Physics Program, Washington State University, Pullman, Washington 99163

J.K. ROY–Department of Chemistry, University of North Carolina, Chapel Hill, North Carolina 27514

B.J. SCHEVE–Chemistry Department, Michigan State University, East Lansing, Michigan 48824

N.J. TURRO–Chemistry Department, Columbia University, New York, New York 10027

P.J. WAGNER–Chemistry Department, Michigan State University, East Lansing, Michigan 48824

R.W. WAKE–Research Laboratories, Eastman Kodak Company, Rochester, New York 14650

A. WELLER–Max-Planck-Institut für Biophysikalische Chemie, Abteilung Spektroskopie, Göttingen, Germany

D.G. WHITTEN–Department of Chemistry, University of North Carolina, Chapel Hill, North Carolina 27514

KLAAS ZACHARIASSE–Max-Planck-Institut für Biophsikalische Chemie, Abteilung Spektroskopie, Göttingen, Germany

# PREFACE

This book consists of the majority of the papers presented at the International Exciplex Conference held at the University of Western Ontario, May 28-31, 1974.

Since the discovery of excimer formation by Förster and Kasper in 1954, the realization of their importance and, subsequently, that of exciplexes, in photochemistry has become widespread. This volume is the first major compilation presenting the various aspects of exciplex behavior and it is hoped that it will provide researchers with a starting point in attempts to understand the present state of knowledge in the area.

The conference owed its success, and even its very existence, to a number of people and organizations. To these, the editors owe a great debt. Included are the U.S. Army Research Office in Durham, North Carolina, which provided the major financial support, the National Research Council of Canada, the Eastman Kodak Company, the GCA/McPherson Company, Photochemical Research Associates and the University of Western Ontario.

In addition, the editors are deeply indebted to two individuals involved in the completion of this manuscript. We thank Jane Hill for completing the artwork not provided by the authors. And words cannot express our gratitude to Cheryl DuCharme whose competence, patience and fortitude enabled her to undertake and complete the onerous task of providing a camera-ready manuscript.

Theodor Förster–1910–1974

This volume is dedicated to the memory of Professor Theodor Förster, who made such important contributions in this field, and who died at the age of 64 on May 20, 1974, shortly before the conference began.

Theodor Förster's academic career began in 1933 when, at 23, he received his doctorate from the University of Frankfort with a thesis entitled "Polarization of Electrons by Reflexion."

He became an assistant of Karl Friedrich Bonhoeffer and went with him one year later to the University of Leipzig which, at that time, with men such as Debye, Heisenberg, Kautsky, and van der Waerden, was one of the most outstanding universities in Germany. There, between the ages of 25 to 30, Theodor Förster carried out theoretical studies on the carbon atom valencies and about the light absorption of organic compounds, which made him well known in the scientific world and a full professor at the age of 32.

During the 5 years after the war, when he was *Abteilungsleiter* at the Max-Planck-Institute for Physical Chemistry at Göttingen, his scientific activity reached a peak: he wrote his book, "Fluoreszenz organischer Verbindungen" and carried out the theoretical and experimental studies on energy transfer which now belong to the classics of photochemistry. As early as 1946, in the first paper on this subject, he pointed out the importance of energy transfer in biological systems, such as in photosynthesis or in cellular nucleic acid material. Also, the first papers on proton transfer reactions of excited molecules appeared during those 5 years after the war, opening up the wide field of investigations on reactions of excited molecules to which an important contribution was made in 1954 when Förster and Kasper discovered what is now called excimer formation.

During the last 5 years he became interested in an up to that time completely neglected aspect of photochemistry: the differentiation between diabatic and adiabatic reactions. He designed and carried out experiments which made a quantitative distinction possible and he worked on the theory which also included radiationless transitions.

Up to his very last day, however, he was still strongly interested in all aspects of the many important fields he had initiated and it was in this spirit that he wrote his contribution to this conference as a survey on excimers and exciplexes covering the more-or-less-well-known features of the subject and forming the basis for the more specialized lectures.

EXCIMERS AND EXCIPLEXES

THE LATE THEODOR FÖRSTER

*Institut für Physikalische Chemie der*

*Universität Stuttgart, Germany*

EXCIMERS

The term "excimer" has become familiar as the name of
atomic or molecular dimer aggregates which are unstable in
their ground state but are stable under electronic excitation.
To be precise one should consider the He dimer as the simplest
excimer and as probably the first one reported in the
literature (1). Emission from an upper $^1\Sigma_u^+$ state to the
dissociative $^1\Sigma_g^+$ ground state produces the Hopfield continuum
which extends from 600 to 1000 $\overset{o}{A}$ in the far UV. This broad
and diffuse emission bears all the essential features which
now are regarded as the characteristics of an excimer.
Similar continua had been observed in the emission spectra
of other rare gases and the vapors of Zn, Hg etc. (2). At
that time, however, vapor-state spectroscopists were not very
interested in diffuse spectra since they do not contain much
information.

Long ago, the formation of excited dimers in solution by
an interaction between excited and unexcited monomers of

aromatic compounds had been considered by Kautsky (3). These dimers were, however, supposed to be nonfluorescent and thought to be responsible merely for the phenomenon of concentration quenching together with some other rather spurious luminescence effects.

In 1954, Kasper and Förster (4) reported the now well-known behaviour of pyrene which, in its more concentrated solutions, exhibits an additional red-shifted structureless fluorescence component without any corresponding feature in the absorption spectrum. Because of the concentration dependence of both spectral components, the red-shifted one was ascribed to a dimeric species (AA)* which does not exist in the ground state but is formed by an excited and an unexcited monomer, both of them in singlet states. The reverse process of dissociation (5) and the various processes of deactivation complete the now-accepted reaction scheme of excimer formation

$$
\begin{array}{ccc}
A^* + A & (AA)^* & \\
\swarrow \searrow & \swarrow \searrow & [1] \\
A+h\nu \quad A & (AA)+h\nu' \quad (AA) &
\end{array}
$$

This mechanism is most clearly demonstrated by the biexponential rise and decay of the dimeric component which Birks and his group were able to measure for the first time (6).

A few years after pyrene, similar concentration changes were reported for 9-alkylanthracenes (7) and for 2,5-diphenyl-oxazole (8). The breakthrough came in 1962 with benzene (9) and naphthalene (10, 11) which then were followed by a multitude of aromatic hydrocarbons and their derivatives (12). As a nonaromatic compound, acetone may be mentioned since a weak excimeric component has been reported in solution fluorescence (13).

At a time when pyrene was still the only well-known example, Stevens and Hutton (14) proposed the name "excimer" which became generally accepted within a few years.

The energetics of excimer formation is best described by a potential energy diagram first used by Stevens and Ban with the interplanar distance of a (most plausible) nearly parallel sandwich configuration as the geometrical coordinate (15). The diagram in figure 1 relates the spectral shift between monomer and excimer band to the stabilization energy of the excimer and to the ground-state repulsion energy.

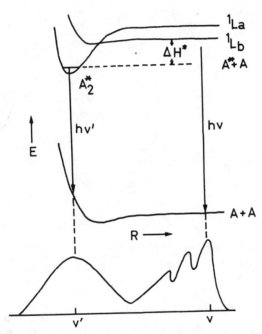

Fig. 1 *Potential energy diagram for photo - association illustrating the origin of excimer and molecular fluorescence bands.*

Because of this repulsion, the stabilization is less than that calculated from the spectral shift alone. Whereas the latter is generally between 5000 and 6000 $cm^{-1}$, corresponding to

14 - 17 kcal/mole, the stabilization energy is merely 10 kcal in the exceptional case of pyrene and much less in most of the others.

Originally, the stability of the pyrene excimer was ascribed to the effect of exciton resonance interaction (16) as a result of the delocalization of the excitation among both components, as

$$A^*A \longleftrightarrow AA^*$$

The excimer state would then result as the lower of the two resonance components which originate from the monomer $^1L_a$ states. The resonance concept seemed to be in accord with the exceptional stability of the pyrene excimer, because of the unusual situation of a strong $^1L_a$ transition at not much higher energy than the emitting $^1L_b$ transition in the monomer.

On the other hand, Ferguson (17) interpreted the electronic state of the excimer as an intermolecular charge-transfer state. This too would have been necessarily formulated as a symmetrical resonance hybrid, as

$$A^+A^- \longleftrightarrow A^-A^+$$

Later, MO theoretical calculations were performed by several authors (18 - 20) under the assumption of a more-or-less symmetrical sandwich geometry of the excimer. Such calculations are not very reliable because of the transannular mode of interaction and because the exact value of the interplanar distance is not experimentally accessible. Nevertheless, it has become clear from these calculations that neither the resonance hypothesis nor the charge-transfer hypothesis alone is sufficient to explain the stability of the excimer state. Instead, exciton resonance and charge-transfer configurations both contribute to the excimer state. In the resonance formulation, this can be described as follows

$$A^*A \longleftrightarrow AA^* \longleftrightarrow A^+A^- \longleftrightarrow A^-A^+$$

This corresponds to the following wave function:

$$\Psi_{exc} = C_1[\Psi(A^*A) + \Psi(AA^*)] + C_2[A^+A^-) + \Psi(A^-A^+)]$$

where the ratio $C_1/C_2$ and with this the relative contributions from exciton and charge-transfer configurations may vary from one case to the other.

Time does not allow me to report all manifestations of excimer formation in detail. I shall only mention excimer formation in crystals (21) where aromatic molecules are arranged in pairs or in other geometries favorable for excimer formation. Imperfections may produce excimer emission in otherwise monomeric emitting crystals (22) or even the reverse (23). Intramolecular excimers are formed in composite molecules where two or more aromatic residues are linked together by flexible $-(CH_2)_n$-chains of suitable length. As Hirayama has demonstrated (24), excimer formation in the diphenylalkane series occurs with n = 3 only. Moreover, in the dicarbazolylalkanes n = 3 does but n = 4 does not (25). Sterically a chain of at least three carbon atoms is necessary for a sandwich conformation to be formed. Longer chains should facilitate monomer emission by the entropy effect of a higher number of competitive conformations. Excimer formation in aromatic polymers such as polystyrene (26, 27) and polyvinylcarbazole (25a) is in accord with this n = 3 rule. On the other hand, excimer emission has been observed in 1,4-naphthylbutane (28) and, very recently, up to n = 8 in the series of (3-pyrenyl)alkanes, $Py-(CH_2)_n-Py$ (29). In the latter case, this might result from the exceptionally strong resonance interaction between pyrene nuclei. In the paracyclophanes, where the phenylene groups are more rigidly connected to each other by two $-(CH_2)_n$-chains, excimer emission has been observed for n = 4 (30).

Excimer formation and subsequent dissociation may contribute to excitation transfer in pure liquid, polymers and other systems with high densities of aromatic components (31, 32). The fluorescence of dinucleotides, polynucleotides and of DNA is of the excimer type (33).

Whereas all our earlier knowledge of excimers had to be drawn from fluorescence spectra, it has now been supplemented by absorption data since the advent of laser flash spectroscopy (34, 35). The respective transitions are, of course, $S_n \leftarrow S_1$ transitions from the (lowest) excimer state to higher excited states.

Interaction of a molecule in its lowest excited singlet state with another one in its ground state is the most common but not the only mechanism of excimer formation. Excimer formation may occur also by a process of triplet-triplet annihilation, as [a]

$$A^\dagger + A^\dagger \rightarrow (AA)^* \qquad [2]$$

and give rise to an excimeric component in delayed fluorescence (36). Another pathway is that of electron transfer between radical anion and radical cation (37), both in their doublet ground states, as

$$A^+ + A^- \rightarrow (AA)^* \qquad [3]$$

Finally, stable dimeric cations may form excimers by electron capture (38):

$$(AA)^+ + e^- \rightarrow (AA)^* \qquad [4]$$

Triplet excimers might be formed either by the interaction of a monomer triplet with a ground-state molecule, as

[a] In this paper triplet states are designated by a dagger (†).

$$A + A \rightarrow (AA) \qquad\qquad [5]$$

or by processes analogous to those in [2] - [4].

They are less stable than singlet excimers because of the smaller magnitude of stabilization by exciton resonance and because of the smaller contribution of charge-transfer configurations. Besides this, they do not form easily under conditions under which they can be observed in phosphorescence emission. Nevertheless, a few trustworthy observations of phosphorescence of inter- as well as of intramolecular excimers do exist (39, 40).

## EXCIPLEXES

Even the proponents of the resonance hypothesis of excimer stability were not surprised by the observation of mixed excimer formation between different but similar molecules such as pyrene and its methyl derivatives (41, 42). It was, however, a surprise when Leonhardt and Weller in 1963 detected an excimer-type emission in mixed solutions of two components so dissimilar as perylene and dimethylaniline (43). Obviously, the stability of such a hetero-excimer could not result from resonance interaction. Consequently, the emitting species was interpreted as a charge-transfer complex $A^- D^+$ which is stabilized by electron transfer from the donor D(dimethylaniline) to the acceptor A (perylene) whereas it is unstable in its ground state AD.

In the meantime formation of hetero-excimers has been observed in numerous systems. It may occur either by an excited acceptor with an unexcited donor or *vice versa* (44)[b]:

[b]  Electronic excitation of the hetero-excimer is indicated solely by the + and - signs of charge separation.

$$A^* + D \rightarrow (A^-D^+) \quad (e.g. \quad A = \text{perylene}, \; D = \text{dimethylaniline})$$

$$D^* + A \rightarrow (D^+A^-) \quad (e.g. \quad D = \text{pyrene}, \; A = 1,4\text{-dicyanobenzene})$$

Usually the component absorbing at higher wavelength is primarily excited but the reverse is also possible (45, 46). As to be expected from the electronic structure of hetero-excimers, these emission spectra are solvent dependent insofar as their maxima are red-shifted with increasing solvent polarity (figure 2). Dipole moments on the order of 10 Debye have been estimated from this (48, 49).

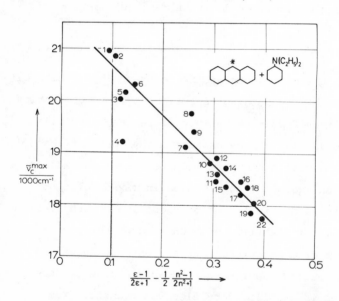

Fig. 2 *Maximum emission frequency of anthracene-diethylaniline exciplex in different solvents (ε = dielectric constant, n = refractive index) ranging from 1,n-hexane to 22, methanol (after Knibbe (47).*

The emission process of a hetero-excimer with pronounced charge-transfer character is

$$(A^-D^+) \rightarrow (AD) + h\nu$$

The emission frequency should, therefore, increase with the ionization potential, I, of the donor and decrease with the electron affinity, E, of the acceptor. Quantitatively the relation can be formulated as

$$h\nu = \varepsilon(A^-D^+) - \varepsilon(AD) = I(D) - E(A) - C \qquad [6]$$

where C represents the coulomb attraction energy at equilibrium distance. As far as this term may be considered constant, linear correlations should be expected in suitable series. Experimentally such linear relations exist (50), (figure 3); however, most of them have slopes less than unity

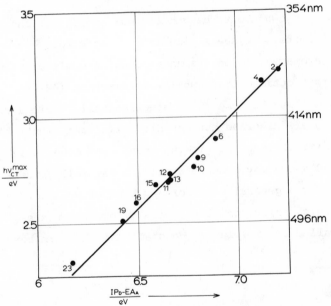

Fig. 3 *Maximum emission frequencies in n-hexane of exciplexes between diethylaniline (D) and aromatic hydrocarbons (A) as a function of I(D) - E(A). Acceptors: 1,naphthalene; 4,phenanthrene; 6,chrysene; 9,1,2-benzpyrene; 10,pyrene; 11,1,2,7,8-dibenzanthracene; 12,1,2,3,4-dibenzanthracene; 13,1,2,5,6-dibenzanthracene; 15,1,2-benzanthracene; 16,anthracene; 19,3,4-benzpyrene; 23,perylene (after Knibbe (47)).*

9

as predicted by equation [6]. Similar correlations exist with polarographic oxidation and reduction potentials (51, 52, 53).

Hetero-excimer emission has been observed in many cases where the direction of charge transfer is less obvious than in the examples already reported. Then a more general resonance formulation of the emitting state should apply which even covers the symmetrical excimer as a limiting case (48)

$$A^-D^+ \longleftrightarrow A^+D^- \longleftrightarrow A^*D \longleftrightarrow AD^*$$

$$\psi_{hetero-exc} = C_1\psi(A^-D^+) + C_2\psi(A^+D^-) + C_3\psi(A^*D) + C_4\psi(AD^*)$$

Not much is known about the geometrical configurations of hetero-excimers. It seems plausible, that hetero-excimers with two aromatic components have sandwich configurations which permit optimum transannular interaction (47).

The ability to complex in electronically excited states is much more widespread between dissimilar components than between similar ones and binding mechanisms other than those considered above may even be present. Naturally, there was a desire for a general name for any kind of well-defined complex which exists in electronically excited states; this is now called an exciplex[c]. We shall use the term here in

[c] This term was first used by Walker, Bednar and Lumry (54) in connection with polar solvent effects on fluorescence spectra of indole where, however, the existence of a well-defined complex as the emitting species may be doubtful. Later, Birks (55) proposed to use the term in the sense defined above and this definition was accepted at the 1968 Chicago Luminescence Conference.

the case of unlike components only.

As has been mentioned before, the spectra of hetero-excimers or exciplexes become red-shifted with increasing solvent polarity. This is paralleled by a strong decrease of the fluorescence quantum yield. Lifetime measurements indicate that this is the combined effect of a decrease in radiative decay and an increase in nonradiative decay (56). The former may result from an increase in heteropolarity of the exciplex in more polar solvents. The main nonradiative process is the formation of a solvated ion radical pair in competition with exciplex formation (57). From transient absorption and photocurrent measurements, this has been demonstrated to occur immediately following electron transfer and prior to solvent relaxation (58, 59). One should keep in mind, however, that solvent relaxation itself is a powerful mechanism for radiationless deactivation.

Without any attempt to be complete I shall mention a few other studies on exciplexes. Formation of intramolecular exciplexes occurs in suitable systems with the donor and acceptor components linked together by an aliphatic chain (60, 61). Hirayama's rule is not too stringent here and exciplex emission has been observed with a chain of even four carbon atoms (62). A peculiar kind of exciplex formation occurs in the symmetrical molecule 9,9'-dianthracene where in polar solvents a charge-transfer configuration $A^+-A^-$ becomes stabilized (63). A triple exciplex ($DD^+A^-$) has been observed (64) with naphthalene (D) and 1,4-dicyanobenzene (A). Triplet exciplexes (65) have been identified in several cases by phosphorescence and by triplet-triplet absorption. Laser emission from intramolecular exciplexes has been reported (66).

Although the most typical exciplexes are those consisting of two aromatic components this is by no means necessary. Alkylamines (44) are suitable donors with aromatic hydrocarbons as acceptors and even fluorescence quenching by dimethylmercury (67) is accompanied by exciplex emission. Exciplex formation in the vapor state of Hg (68) and Cd (69) in their $^3P_o$ states has been observed with $NH_3$.

Further details on the spectroscopic properties of excimers and exciplexes are collected in several review articles (70 - 73). The remainder of this chapter shall be devoted to the possible role of excimers and exciplexes in photochemistry.

PHOTOCHEMICAL ASPECTS

It is not surprising that photochemists have been attracted by the concepts of excimers and exciplexes. Electronically excited molecules have been observed as the final products of photochemical reactions in several cases (74 - 76), even in condensed phases. The participation of excimers or exciplexes as electronically excited intermediates in photodimerizations and photocondensations seems therefore quite plausible.

Actually excimers were introduced into photochemistry even before they had been spectroscopically detected. In the formulation of his mechanism of anthracene dimerization, Weigert (77) (1928) deduced the existence of an intermediate which Suzuki (78) (1950) later interpreted as an excited species. In 1952, still two years before the detection of excimer fluorescence the present author (79) formulated the mechanism of anthracene dimerization as follows:

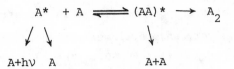

The participation of an excimer in the photodimerization of anthracene and its derivatives has become even more plausible by later observations that, in the amorphous state at lower temperatures, anthracene exhibits excimer fluorescence (80) and that some derivatives even do so in solutions where dimerization takes place (81).

From more recent investigations, it is still not yet certain whether the excited intermediate in photoassociation is identical with the excimer (82, 83). In an investigation of the analogous photodimerization of 2-methoxynaphthalene, where excimer emission could be followed as well, the authors are reluctant to consider the excimer as an intermediate (84).

More convincing, even if only qualitative, evidence for an excimer intermediate is given by the photoreversible cyclomerization of 1,3-bis(α-naphthyl)propane where the reverse reaction is accompanied by excimer fluorescence (85). In several other photodimerizations and photocylomerizations which occur from singlet states, excimers have been suggested as plausible intermediates even if - as in the case of anthracene - no excimer emission has been observed under reaction conditions. The (unsensitized) photodimerization of acenaphthylene (86) may be mentioned as an example.

The cyclomerization of 6,6'-trimethylenebis(4-methyl-pyrone) (87) is an example of an intramolecular photoreaction in which Hirayama's rule is strictly followed. There are, however, other reactions which do not obey this rule. N,N'-alkylenebismaleides cyclomerize unsensitized with moderate

quantum yields for up to six carbon atoms in the chain
between the unsaturated components (88). In the series of
biscoumarins, exclusively *endo* cyclomerization occurs in the
singlet reaction up to n = 11 (88, 89). The *endo* product is
that one would expect to be formed *via* an excimer as intermediate.

The deviations from the n = 3 rule in photoreactions
should not be regarded as evidence against an excimer as a
kinetic intermediate in the photocyclomerization. Generally,
fluorescence in composite systems occurs under conditions of
equilibrium between the different conformations and the
absence of excimer fluorescence for n > 3 has been interpreted
as an entropy effect. In the case of a photochemical reaction
where no fluorescence occurs such an equilibrium does not
necessarily exist so that kinetic rather than equilibrium
conditions prevail. Recent results obtained with a dicoumarin
point to some participation of ground-state interaction (90).

Another example where strong arguments have been
obtained for an excimer as intermediate is the photodimerization of thymine which is supposed to play an important role
in the protection of DNA against photochemical damage (91, 92).
A role similar to that of excimers in photodimerizations may
be expected for exciplexes in photoadditions; only one
example is mentioned here. In mixed solutions of 9,10-dimethyl-
anthracene and 9-cyanoanthracene, the crossed photodimer is
preferentially formed and the dependence of its yield on
solvent polarity is that to be expected from the variation in
stability of the respective exciplexes (93).

Although the participation of excimers and of exciplexes
as intermediates in photoreactions of the kind mentioned
before seems likely, it cannot yet be considered as proven.
Even if excimer or exciplex emission occurs under reaction

conditions, only quantitative spectral and photochemical data under varying conditions and the representation of both by a common reaction mechanism may be considered as a proof. Nevertheless, the assumption of such intermediates may be useful as a heuristic principle for photochemical work.

## REFERENCES

1. J.J. Hopfield, *Phys. Rev.* 35, 1133 (1930); 36, 784 (1930).

2. W. Finkelnburg, "Kontinuierliche Spektren." Springer, Berlin, 1938.

3. H. Kautsky and H. Merkel, *Naturwiss.* 27, 195 (1939).

4. Th. Förster and K. Kasper, *Z. Phys. Chem.*, N.F. 1, 275 (1954).

5. E. Döller and Th. Förster, *Z. Phys. Chem.*, N.F. 34, 132 (1962).

6. J.B. Birks, D.J. Dyson and I.H. Munro, *Proc. Roy. Soc. (London)*, A, 275, 575 (1963).

7. G.A. Tischenko, B. Ya. Sveshnikov and A.S. Cherkasov, *Optika i Spektroskopija* 4, 631 (1958).

8. I.B. Berlman, *J. Chem. Phys.* 34, 1083 (1961).

9. T.V. Ivanova, G.A. Mokeeva and B. Ya. Sveshnikov, *Optika i Spektroskopija* 12, 586 (1962); *Optics and Spectroscopy* 12, 325 (1962).

10. E. Döller and Th. Förster, *Z. Phys. Chem.*, N.F. 31, 274 (1962).

11. I.B. Berlman and A. Weinreb, *Mol. Phys.* 5, 313 (1962).

12. J.B. Birks and L.G. Christophorou, *Proc. Roy. Soc. (London)*, A, 274, 552 (1963).

13. M. O'Sullivan and A.C. Testa, *J. Amer. Chem. Soc.* 90, 6245 (1968).

14. B. Stevens and E. Hutton, *Nature* 186, 1045 (1960).

15. B. Stevens and M.I. Ban, *Trans. Faraday Soc.* 60, 1515 (1964).

16. Th. Förster and K. Kasper, *Z. Elektrochem.* 59, 976 (1955).

17. J. Ferguson, *J. Chem. Phys.* *28*, 765 (1958).

18. E. Konijnenberg, Dissertation, Vrije Universiteit te Amsterdam (1963).

19. J.N. Murrell and J. Tanaka, *Mol. Phys.* *7*, 363 (1964).

20. T. Azumi and S.P. McGlynn, *J. Chem. Phys.* *42*, 1675 (1965).

21. B. Stevens, *Spectrochim. Acta* *18*, 439 (1962).

22. H.W. Offen, *J. Chem. Phys.* *44*,699 (1966).

23. N.Y.C. Chu and D.R. Kearns, *Molec. Cryst. Liq. Cryst.* *16*, 61 (1972).

24. F. Hirayama, *J. Chem. Phys.* *42*, 3163 (1965).

25. W. Klöpffer, *Chem. Phys. Letters* *4*, 193 (1969).

25a. W. Klopffer, *J. Chem. Phys.* *50*, 2337 (1969).

26. L.J. Basile, *J. Chem. Phys.* *36*, 2204 (1962).

27. S.S. Yaneri, F.A. Bovey and R. Lumry, *Nature* *200*, 242 (1963).

28. E.A. Chandross and C.J. Dempster, *J. Amer. Chem. Soc.* *92*, 3586 (1970).

29. W. Kühnle and K. Zachariasse, personal communication.

30. M.T. Vala, J. Haebig and S.A. Rice, *J. Chem. Phys.* *43*, 886 (1965).

31. J.B. Birks and J.C. Conte, *Proc. Roy. Soc. (London) A, 303*, 85 (1968).

32. J. Klein and R. Voltz, *J. Chim. Phys. Physicochim. Biol* *67*, 704 (1970).

33. J. Eisinger, M. Gueron, R.G. Schulman and T. Yamane, *Proc. Natl. Acad. Sci. U.S.* *55*, 1015 (1966).

34. C.R. Goldschmidt and M. Ottolenghi, *J. Phys. Chem.* *74*, 2041 (1970).

35. M.A. Slifkin and A.O. Al-Chalabi, *Chem. Phys. Letters* *20*, 211 (1973).

36. C.A. Parker and C.G. Hatchard, *Nature* *190*, 165 (1961).

37. E.A. Chandross, J.W. Longworth and R.E. Visco, *J. Amer. Chem. Soc.* *87*, 3259 (1965).

38. B. Brocklehurst and R.D. Russell, *Nature* *65*, 213 (1967).

39. J. Langelaar, R.P.H. Rettschnick, A.M.F. Lambooy and G.J. Hoytink, *Chem. Phys. Letters* *1*, 609 (1968).

40. P. Avouris and M.A. El-Bayoumi, *ibid.* *20*, 59 (1973).

41. J.B. Birks and L.G. Christophorou, *Nature* *196*, 33 (1962).

42. B.K. Selinger, *ibid.* 203, 1062 (1964).

43. H. Leonhardt and A. Weller, *Ber. Bunsenges. Phys. Chem.* *67*, 791 (1963).

44. H. Knibbe and A. Weller, *Z. Phys. Chem.*, *N.F.* *56*, 99 (1967).

45. R.J. McDonald and B.K. Selinger, *Aust. J. Chem.* *24*, 1795 (1971).

46. A.E.W. Knight and B.K. Selinger, *Chem. Phys. Letters* *10*, 43 (1971).

47. H. Knibbe, Dissertation, Vrije Universiteit te Amsterdam (1969).

48. H. Beens and A. Weller, *Acta Phys. Polon* *34*, 593 (1968).

49. H. Beens, H. Knibbe and A. Weller, *J. Chem. Phys.* *47*, 1183 (1967).

50. H. Knibbe, D. Rehm and A. Weller, *Z. Phys. Chem.*, *N.F.* *56*, 95 (1967).

51. D. Rehm and A. Weller, *Z. Phys. Chem.*, *N.F.* *69*, 183 (1970).

52. N. Mataga and K. Ezumi, *Bull. Chem. Soc. Japan* *40*, 1355 (1967).

53. H. Knibbe, D. Rehm and A. Weller, *Ber. Bunsenges. Phys. Chem.* *73*, 839 (1969).

54. M.S. Walker, T.W. Bednar and R. Lumry, *J. Chem. Phys.* *45*,

3455 (1966); *47,* 1020 (1967).

55. J.B. Birks, *Nature 214,* 1187 (1967).

56. N. Mataga, T. Okada and N. Yamamoto, *Chem. Phys. Letters 1,* 119 (1967).

57. H. Knibbe, K. Röllig, F.P. Schäfer and A. Weller, *J. Chem. Phys. 47,* 1184 (1967).

58. C.R. Goldschmidt, R. Potashnik and O. Ottolenghi, *J. Phys. Chem. 75,* 1025 (1971).

59. Y. Taniguchi and N. Mataga, *Chem. Phys. Letters 13,* 596 (1972).

60. E.A. Chandross and H.T. Thomas, *Chem. Phys. Letters 9,* 393 (1971).

61. T. Okada, T. Fujita, M. Kubota, S. Masaki, N. Mataga, R. Ide, Y. Sakata and S. Misumi, *Chem. Phys. Letters 14,* 563 (1972).

62. G.S. Beddard, R. Davidson and A. Lewis, *J. Photochem. 1,* 491 (1973).

63. F. Schneider and E. Lippert, *Ber. Bunsenges. Phys. Chem. 72,* 1155 (1968).

64. H. Beens and A. Weller, *Chem. Phys. Letters 2,* 140 (1968).

65. G. Briegleb, H. Schuster and W. Herre, *Chem. Phys. Letters 4,* 53 (1969).

66. N. Nakashima, N. Mataga, C. Yamanaka, R. Ide and S. Misumi, *Chem. Phys. Letters 18,* 386 (1973).

67. E. Van der Conokt, D. Lietaer and M. Matagne, *J. Chem. Soc.* Faraday Trans. 2, 69, 322 (1973).

68. A.B. Callear and J.H. Connor, *Chem. Phys. Letters 13,* 245 (1972); *14,* 384 (1972).

69. P.D. Morten, C.G. Freeman, M.J. McEwan, R.F.C. Claridge and L.F. Phillips, *Chem. Phys. Letters 16,* 148 (1972).

70. Th. Förster, *Angew. Chem. 81,* 364 (1969).

71. B. Stevens, *Advances in Photochemistry* *8*, 161 (1971).

72. J.B. Birks *in* "Progress in Reaction Kinetics," (G. Porter, ed.), Vol. 5, p. 181. Pergamon Press, London, 1970.

73. W. Klöpffer *in* "Organic Molecular Photophysics," (J.B. Birks, ed.), p. 357. J. Wiley and Sons, New York, 1973.

74. Th. Förster, *Pure Appl. Chem.* *24*, 443 (1970).

75. J. Menter and Th. Förster, *Photochem. Photobiol.* *15*, 289 (1972).

76. N.J. Turro, P. Lechtken, A. Lyons, R.R. Hautala, E. Carnahan and T.J. Katz, *J. Amer. Chem. Soc.* *95*, 2035 (1973).

77. F. Weigert, *Naturwiss.* *15*, 124 (1927).

78. M. Suzuki, *Bull. Chem. Soc. Japan* *23*, 120 (1950).

79. Th. Förster, *Z. Elektrochem.*, *Ber. Bunsenges. Phys. Chem.* *56*, 716 (1952).

80. H.H. Perkampus and L. Pohl, *Z. Phys. Chem. N.F.*, *40*, 162 (1964).

81. A.S. Cherkasov and T.M. Vember, *Optika i Spektroskopija* *6*, 503 (1959); *Optics and Spectroscopy* *6*, 319 (1959).

82. T.M. Vember, T.V. Veselova, I.E. Obyknovenaga, A.S. Cherkasov and V.I. Shirokov, *Izv. Akad. Nauk SSR, Ser. Fiz.* *37*, 837 (1973).

83. J. Menter, unpublished results.

84. P. Wilairat and B.K. Selinger, *Aust. J. Chem.* *21*, 733 (1968).

85. E.A. Chandross and C.J. Dempster, *J. Amer. Chem. Soc.* *92*, 703, 704, 3586 (1970).

86. I.M. Hartmann, W. Hartmann and G.O. Schenck, *Chem. Ber.* *100*, 3146 (1967).

87. R.A. Keller, *J. Amer. Chem. Soc.* *90*, 1940 (1968).

88. J. Put and F.C. DeSchryver, *J. Amer. Chem. Soc.* *95*, 137 (1973).

89. L. Leenders and F.C. DeSchryver, in press.

90. F.C. DeSchryver, personal communication.

91. J. Eisinger, *Photochem. Photobiol.* *7*, 597 (1968).

92. A.A. Lamola and J. Eisinger, *Proc. Nat. Acad. Sci. U.S.* *59*, 4 (1968).

93. H. Bouas-Laurent and R. Lapouyade, *Chem. Commun.*, 817 (1969).

# SINGLET-AND TRIPLET-STATE EXCIPLEXES

A. WELLER

*Max-Planck-Institut für biophysikalische Chemie,*

*Abteilung Spektroskopie, Göttingen, Germany*

## CLASSIFICATION OF EXCIPLEXES

In 1963, when hetero-excimer formation was reported for the first time and the new emitting species was interpreted as an excited charge-transfer complex state (1), it was pointed out that the excited molecular complexes which are formed in the excited singlet state by a one-to-one association of electron donor (D) and acceptor (A) molecules do not differ in principle from excited stable ground-state complexes [*cf.* (2)].

The ground-state stability of these EDA complexes is ascribed to the interaction (3)

$$^1(AD) \xrightleftharpoons{S^*_{ad}} {}^1(A^-D^+)$$

I

between the no-bond ground state and the polar (charge-transfer) excited state. This interaction, which depends on the overlap integral $S^*_{ad}$ between the lowest antibonding orbital (a$^*$) of the acceptor and the highest bonding orbital (d) of the donor, also exerts, of course, a destabilizing effect on the excited (charge-transfer) state amounting to (4)

$$U_{dest} = \frac{(\alpha - E^o_{CT} S^*_{ad})^2}{E^o_{CT}}$$

[1]

where $\alpha$ is the hamiltonian matrix element between the zero-order complex ground and excited states and $E^o_{CT}$ is the energy of the "pure" charge-transfer state

$$^1(A^-D^+)$$

II

which has a maximum dipole moment of about 14 Debye (5,6).

On the other hand, as the charge-transfer complex state in most cases has an energy close to that of the lowest excited singlet states of the components, it is clear that a more general description of the emitting exciplex state should also include locally excited states (5, 6, 9). Their inter-actions with the charge-transfer state

$$^1(\overset{*}{AD}) \xleftrightarrow{\;S_{ad}\;} {}^1(A^-D^+) \xleftrightarrow{\;S^{**}_{ad}\;} {}^1(A\overset{*}{D})$$

III

depend upon the overlap integrals $S_{ad}$, between the highest bonding orbitals, and $S^{**}_{ad}$, between the lowest antibonding orbitals, respectively, and lead to a stabilization of the emitting exciplex state by an amount $U_{stab}$ which depends on the overlap integrals and increases as the charge-transfer state comes closer in energy to the locally excited states.

These considerations apply first of all to the gas phase where the energy of the "pure" charge-transfer exciplex state above the separated ground-state components is given by

$$E^o_{CT} = IP_D - EA_A - C$$

[2]

where $IP_D$ is the ionization potential of the donor, $EA_A$ is the electron affinity of the acceptor and $C$ is the coulomb attraction energy at equilibrium distance.

Taking into account the aforementioned interactions as well as possible solvation effects one obtains, quite generally, for the energy, $E_e$, (again relative to the separated ground-state components) of any exciplex in any solvent

$$E_e = E_{CT}^o + U_{dest} - U_{stab} - \Delta H_e^{sol} \qquad [3]$$

The enthalpy of solvation, $\Delta H_e^{sol}$, is associated with the equilibrium

$$A_g + D_g + (A^-D^+)_s \rightleftharpoons A_s + D_s + (A^-D^+)_g \qquad [4]$$

where the subscripts g and s refer to gas phase and solution respectively. Since the solvation of the exciplex (with dipole moment $\mu$) is by far the dominating factor, one can write

$$\Delta H_e^{sol} = \frac{\mu^2}{\rho^3} (\frac{\varepsilon - 1}{2\varepsilon + 1} - \frac{d \ln \varepsilon}{d \ln T} \frac{3\varepsilon}{(2\varepsilon + 1)^2}) \qquad [5]$$

where $\rho$ is the equivalent sphere radius in the Kirkwood-Onsager continuum model, which is applied here, and $\varepsilon$ is the dielectric constant.

It is on the basis of equation [3] that the following three types of exciplexes, discussed in the foregoing section, can be distinguished:

Excited EDA complexes (type I) have
$$U_{dest} > U_{stab} \approx 0 \text{ so that } E_e > E_{CT}^o - \Delta H_e^{sol}$$

Charge-transfer exciplexes (type II) have
$$U_{dest} \approx U_{stab} \approx 0 \text{ so that } E_e \approx E_{CT}^o - \Delta H_e^{sol}$$

25

Mixed excimers and excimers (type III) have

$$U_{stab} > U_{dest} \approx 0 \quad \text{so that} \quad E_e < E_{CT}^o - \Delta H_e^{sol}$$

This classification which has been formulated for singlet-state exciplexes (6-8) can also be applied to triplet-state exciplexes if account is taken of the fact that in this case $U_{dest} = 0$ because of the difference in multiplicity between the no-bond ground and charge-transfer excited states.

The experimental verification of this exciplex classification has been found to be considerably facilitated by using, instead of $IP_D$ and $EA_A$ in equation [2], the polarographic half-wave oxidation and reduction potentials $E_D^{ox}$ and $E_A^{red}$, respectively, which are linearly related to $IP_D$ and $EA_A$ (and much more easily accessible). This replacement yields

$$E_{CT}^o = E_D^{ox} - E_A^{red} + (\Delta G_{D+}^{sol} + \Delta G_{A-}^{sol} - C) \qquad [6]$$

where $\Delta G_{D+}^{sol}$ and $\Delta G_{A-}^{sol}$ are the solvation free energies associated with the processes

$$D_g + D_s^+ \rightleftharpoons D_s + D_g^+ \qquad [7]$$

and

$$A_g + A_s^- \rightleftharpoons A_s + A_g^- \qquad [8]$$

where the subscript g refers to the gas phase and s to the (polar) solvent in which the oxidation and reduction potentials have been determined.

SINGLET-STATE EXCIPLEXES

The advantage of using one-electron oxidation and reduction potentials instead of ionization potentials and electron affinities has recently been demonstrated (7, 11) when it was found that the maximum fluorescence emission energies, $h\nu_e^{max}$, in *n*-hexane of more than 160 typical hetero-excimer systems in the range between 1.9 and 3.5 e$V$ (650 - 350 nm) follow the relation

$$h\nu_e^{max} \text{ (hex)} = E_D^{ox} - E_A^{red} - 0.15 \pm 0.10 \text{ } eV \qquad [9]$$

and thus fall between the two dashed lines of figure 1. The unit slope obtained here also holds for equation [6] (see below) and is, evidently, due to the fact that the solvation free energies $\Delta G_{D+}^{sol} + \Delta G_{A-}^{sol}$ of the separate ions not only are of the same order of magnitude as the Coulomb attraction energy term but also depend in the same way on the size of the ions as does C.

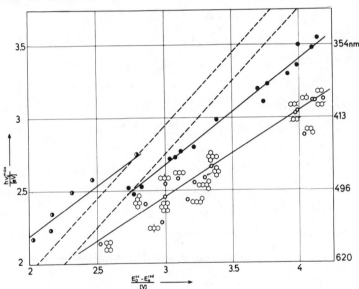

Fig. 1 Plot of maximum energy of exciplex fluorescence $h\nu_e^{max}$, measured in n-hexane at room temperature, against the difference between donor oxidation and acceptor reduction potentials, $E_D^{ox} - E_A^{red}$. The two dashed lines give the limits for charge-transfer singlet exciplexes according to equation [9].

◐ Excited EDA complexes; ● mixed excimers;

○ excimers.

The maximum emission energies of some typical EDA complexes in inert solvents are presented in figure 1 by the half-filled circles. As expected for type I complexes, all these points lie above the upper limit given by equation [9]

Also in accordance with the classification given above is the finding that the maximum emission energies of excimers (*cf.* circles in figure 1) are considerably lower than one would expect if only charge-transfer interaction was present and thus they fall below the lower dashed line in figure 1. Complex systems like those represented by the filled circles, which are in between charge-transfer exciplexes and excimers have been called mixed excimers (6). Their intermediate behaviour also manifests itself in that they invariably have dipole moments smaller than 14 Debye and are formed (and emit with reasonable quantum yield) even in strongly polar solvents such as acetonitrile (6, 9). Another characteristic of mixed excimers seems to be that either $E_A^{ox} - E_D^{ox}$ or $E_A^{red} - E_D^{red}$ are relatively small which means that there is no marked difference in the donor or acceptor properties of the complex components (*cf.* table 1). For excimers (A = D) these differences are, of course, zero.

As the exciplex classification given above applies directly to exciplex energies rather than to their emission maxima, we have determined the energies, $E_e$ of more than 60 exciplex systems in aliphatic hydrocarbon solvents according to

$$E_e = h\nu_{o,o} - \Delta H_d \tag{10}$$

by subtracting from the molecular zero-zero fluorescence energy, $h\nu_{o,o}$, the exciplex dissociation enthalpy, $\Delta H_d$, which is obtained from the temperature dependence (van 't Hoff plot) of the relative exciplex/molecular fluorescence intensity

## TABLE 1

*Data on mixed excimers and excimers*

| No | Acceptor $(-E^{red})$[a] | Donor $(E^{ox})$[a] | $E_e$ (eV) | $U_{stab}$ (eV) | $E_A^{ox}-E_D^{ox}$ (V) | $E_A^{red} - E_D^{red}$ (V) |
|---|---|---|---|---|---|---|
| 1 | Benz(c)acridine*(1.75 $V$) | 1,2,4-(MeO)$_3$C$_6$H$_3$(1.12 $V$) | 2.84 | 0.25 | 0.07 | > 1.5 |
| 2 | Benz(a)anthracene(2.02 $V$) | 9,10-Me$_2$-anthracene*(0.94 $V$) | 2.87 | 0.31 | 0.23 | 0.01 |
| 3 | Anthracene*(1.96 $V$) | 1,2,4-(MeO)$_3$C$_6$H$_3$(1.12 $V$) | 2.94 | 0.36 | 0.02 | > 1.3 |
| 4 | Benz(a)anthracene*(2.02 $V$) | 1,2,4-(MeO)$_3$C$_6$H$_3$(1.12 $V$) | 2.97 | 0.39 | 0.05 | > 1.2 |
| 5 | Pyrene*(2.10 $V$) | 1,2,4-(MeO)$_3$C$_6$H$_3$(1.12 $V$) | 3.04 | 0.40 | 0.08 | > 1.2 |
| 6 | Pyrene*(2.10 $V$) | Pyrene(1.20 $V$) | 2.90 | 0.72 | 0 | 0 |
| 7 | Naphthalene*(2.58 $V$) | Naphthalene(1.60 $V$) | 3.69 | 0.81 | 0 | 0 |

[a] measured in acetonitrile or dimethylformamide against S.C.E.

* asterisk denotes primarily excited species.

(10, 12). The results plotted in figure 2 show that most of the exciplexes investigated obey the relation

$$E_e = E_D^{ox} - E_A^{red} + 0.15 \pm 0.10 \text{ } eV \qquad [11]$$

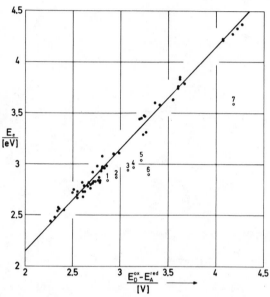

Fig. 2 *Plot of singlet exciplex energy, $E_e$, in aliphatic hydrocarbon solvents against the difference between donor oxidation and acceptor reduction potentials, $E_D^{ox} - E_A^{red}$.*
*Equation of the line: $E_e = E_D^{ox} - E_A^{red} + 0.15$ eV*
*1 - 5, mixed excimers; 6, pyrene excimer; 7, naphthalene excimer.*

represented by the straight line. Evidently these systems are type II charge-transfer exciplexes for which $U_{dest} - U_{stab} \approx 0$ so that with equations [3], [6] and [9] one obtains

$$E_{CT}^o - \Delta H_e^{sol} = E_D^{ox} - E_A^{red} + 0.15 \pm 0.10 \text{ } eV \qquad [12]$$

Calculations, using equation [5], of the solvation enthalpy term in aliphatic hydrocarbon solvents yield, with $\mu^2/\rho^3 = 0.7 \pm 0.3$ eV (5, 6), $\Delta H_e^{sol} = 0.17 \pm 0.07$ eV so that

$$E_{CT}^{o} = E_{D}^{ox} - E_{A}^{red} + 0.32 \text{ e}V \qquad [13]$$

if one assumes that the experimental error limits of $\pm$ 0.10 eV are mainly caused by variations in the solvation enthalpy.

Equation [13] gives the energy of the unperturbed charge-transfer state in the gas phase in terms of the oxidation and reduction potentials of the molecular components. The more general expression, obtained with equation [3],

$$E_{e} = E_{D}^{ox} - E_{A}^{red} + U_{dest} - U_{stab} - \Delta H_{e}^{sol} + 0.32 \text{ e}V \quad [14]$$

can be used together with equation [5] to calculate the stabilization energies, $U_{stab}$, of excimers and mixed excimers (see below) or to make very reliable estimates of the energies of charge-transfer exciplexes (where $U_{dest} - U_{stab} \approx 0$) in any solvent if the dipolar energy $(\mu^2/\rho^3)$ has been obtained from the red shift of the exciplex emission band with increasing solvent polarity (5, 9).

The numbered systems in figure 2 which clearly deviate from charge-transfer exciplex behaviour are listed in table 1. This includes the pyrene and naphthalene excimers (13, 14).

The stabilization energies, $U_{stab}$, of the mixed excimers (systems 1 - 5) and excimers (6 and 7) have been calculated with the aid of equation [14] with $U_{dest} = 0$ and $\Delta H_{e}^{sol} = 0.10$ eV $(\mu^2/\rho^3 \approx 0.4$ eV) for mixed excimers and $\Delta H_{e}^{sol} = 0$ for excimers. It is interesting to point out that the value of $E_{CT}^{o}$ calculated according to equation [13] comes close to or is even greater than the lowest $L_a$-state energy of the components and that the stabilization energy of mixed excimers is just about half that of the excimers.

## TRIPLET-STATE EXCIPLEXES

As the two unpaired electrons in a typical charge-transfer exciplex occupy different orbitals which are localized in the complex components, the singlet-triplet splitting in the charge-transfer state is expected to be small and, in fact, will be zero for the zero-order charge-transfer state. It is, therefore, to be expected that equations [2], [6] and [13] in the form

$$E_{CT}^{o} = E_D^{ox} - E_A^{red} + \Delta^o \qquad [15]$$

will be applicable also to triplet states.

Generally phosphorescence is observable only in rigid media where diffusion is extremely slow. Hence exciplex phosphorescence can be studied only with complexes which are present in the ground state, and it is under these conditions that exciplex phosphorescence spectra, indeed, have been observed (15, 16). Their peak energies, $h\nu_e^{max}(ph)$, obtained in rigid glass solution at 77 K are plotted in figure 3 against the difference of the donor oxidation and the acceptor reduction potentials. The two dashed lines, as in figure 1, correspond to the limits given in equation [9]. Clearly, in exciplex phosphorescence, deviations from unit slope behaviour are more frequent than in exciplex fluorescence. Nevertheless, there is still some regularity and most deviations can be rationalized. Strong positive deviations, as observed with systems represented by filled circles, occur typically at low values of $E_D^{ox} - E_A^{red}$, i.e. with strongly bound EDA complexes, and reflect the contribution of the ground-state binding energy, $\Delta H_g$, to the exciplex emission (cf. equation [21]). Systems represented by circles and showing negative deviations have a phosphorescence spectrum similar to but

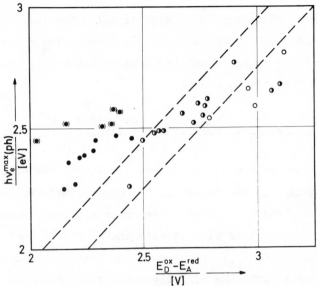

Fig. 3 Plot of maximum energy of exciplex phosphorescence, $h\nu_e^{max}$ (ph), measured in rigid glass solution at 77 K against the difference between donor oxidation and acceptor reduction potentials, $E_D^{ox} - E_A^{red}$. The two dashed lines give the limits found for charge-transfer singlet-state exciplexes (cf. equation [9]).

much less structured than the phosphorescence spectrum of the acceptor. The emitting state in these cases must be considered as a slightly perturbed locally excited (acceptor) triplet state $^3$(AD)*.

Similar to exciplex fluorescence, phosphorescence from triplet-state exciplexes, in order to be observable as a new emission, requires the zero-order energy of the unperturbed charge-transfer triplet state, $^3(A^-D^+)_o$, to be below or at least not too much above that of the lowest locally excited triplet state, $^3$(AD)$^*_\ell$, of the complex, so that the interaction

$$^3(A^-D^+)_o \longleftrightarrow \, ^3(AD)^*_\ell$$

IV

can produce an emitting state which is below the lowest
triplet state of the components. This interaction causes the
triplet exciplex to be stabilized (relative to the unperturbed
charge-transfer triplet state) by

$$U_{stab} = \left( \frac{(h\nu_{o,o}(ph) - E_{CT}^{o})^2}{4} + \beta^2 \right)^{\frac{1}{2}} - \frac{h\nu_{o,o}(ph) - E_{CT}^{o}}{2} \qquad [16]$$

where $h\nu_{o,o}(ph)$ is the energy of the lowest triplet state of
the components, $E_{CT}^{o}$ the energy of the unperturbed charge-
transfer state and $\beta$ the hamiltonian matrix element between
them. This interaction also affects the degree of charge-
transfer character, $\gamma_{CT}$, of the emitting triplet state that
can be estimated from the zero-field-splitting parameters
obtained by ESR spectroscopy (17 - 19). Since $^3(AD)_{\ell}^{*}$ and
$^3(A^-D^+)_{o}$ with the energies $h\nu_{o,o}(ph)$ and $E_{CT}^{o}$, respectively,
are the two interacting states, one has

$$\gamma_{CT} \gtreqless 0.5 \qquad [17]$$

if

$$h\nu_{o,o}(ph) \gtreqless E_{CT}^{o} \qquad [18]$$

or, with equation [15], if

$$h\nu_{o,o}(ph) - (E_{D}^{ox} - E_{A}^{red}) \gtreqless \Delta^{o} \qquad [19]$$

This leads, with the available $\gamma_{CT}$ and energy data collected
in (19), to maximum ($< 0.34$ eV) and minimum ($> 0.31$ eV) values
for $\Delta^{o}$ which are in surprisingly good agreement with

$$\Delta^{o} = 0.32 \text{ eV}$$

and thus confirm the assumption that the zero-order charge-
transfer singlet and triplet-states are degenerate.

Since $U_{dest} = 0$ (see above), one has for the triplet state
exciplex energy in solution

$$^3E_e = E_{D}^{ox} - E_{A}^{red} - U_{stab} - \Delta H_{e}^{sol} + 0.32 \text{ eV} \qquad [20]$$

where $U_{stab}$ is given by equation [16] and $\Delta H_e^{sol}$ has the same
meaning as in equation [3]. For the peak energy of the
exciplex phosphorescence one can write

$$h\nu_e^{max}(ph) = E_D^{ox} - E_A^{red} - U_{stab} + \Delta H_g - \Delta E + 0.32 \text{ eV} \quad [21]$$

where $\Delta E$ comprises contributions due to solvation and Franck-
Condon effects which together may cause a red shift of
approximately 0.5 eV.

When equation [20] is applied to excimers (where
$\Delta H_e^{sol} = 0$), one finds that even the existence of only slightly
stable triplet excimers is highly improbable. In the case of
pyrene, where $E_{CT}^o = 3.62$ eV and $h\nu_{o,o}(ph) = 2.09$ eV, a
stabilization energy $U_{stab} > 1.53$ eV would be necessary; one
can estimate that, according to equation [16], the interaction
matrix element would have to have the impossibly high value
of $|\beta| \geq 0.56$ eV in order to produce a triplet excimer which
at, say, 150 K is stable enough against dissociation into
triplet and ground-state monomer to live for about 1 μsec.

Another application of equation [20] together with
equation [14] refers to the exciplex systems which are
represented in figure 3 by the points in parentheses. In
these systems the delayed emission spectrum is identical with
the exciplex fluorescence spectrum and is, therefore, assumed
to be an E-type delayed fluorescence (20) from the exciplex
singlet state which, even at 77 K, is thermally populated
from the exciplex triplet state. This assumption is in
agreement with the observation that the decay times are
substantially shorter than those of the other exciplexes.
Comparison of equations [14] and [20] shows that the singlet-
triplet splitting which is of the order of 0.1 eV (21) is
approximately given by $U_{dest}$ . It is with these exciplex

systems that exciplex formation in the triplet state followed by E-type delayed fluorescence can be studied in fluid solution at room temperature (16).

## REFERENCES

1. H. Leonhardt and A. Weller, *Ber. Bunsenges. Phys. Chem.* <u>67</u>, 791 (1963).

2. G. Briegleb, "Elektronen-Donator-Acceptor-Komplexe." Springer-Verlag, Heidelberg, 1961.

3. R.S. Mulliken, *J. Amer. Chem. Soc.* <u>72</u>, 600 (1950); *74*, 811 (1952); *J. Phys. Chem.* <u>56</u>, 801 (1952).

4. H. Beens, Ph.D. Thesis, Free University, Amsterdam (1969).

5. H. Beens, H. Knibbe and A. Weller, *J. Chem. Phys.* <u>47</u>, 1183 (1967).

6. H. Beens and A. Weller, *Acta Phys. Polon.* <u>34</u>, 593 (1968).

7. D. Rehm and A. Weller, *Z. Phys. Chem. N.F.* <u>69</u>, 183 (1970).

8. N. Mataga, T. Okada and K. Ezumi, *Mol. Phys.* <u>10</u>, 203 (1966).

9. H. Knibbe, Ph.D. Thesis, Free University, Amsterdam (1969).

10. H. Knibbe, D. Rehm and A. Weller, *Ber. Bunsenges. Phys. Chem.* <u>73</u>, 839 (1969).

11. D. Rehm, *Z. Naturforsch.* <u>25a</u>, 1442 (1970).

12. J. Guttenplan, D. Rehm and A. Weller, unpublished results.

13. Th. Förster and H.P. Seidel, *Z. Phys. Chem. N.F.* <u>45</u>, 58 (1965).

14. J.B. Aladekomo and J.B. Birks, *Proc. Roy. Soc. (London)* A <u>284</u>, 551 (1965).

15. S. Iwata, J. Tanaka and S. Nagakura, *J. Chem. Phys.* <u>47</u>, 2203 (1967).

16. H. Beens and A. Weller *in* "Molecular Luminescence" (E.C. Lim, ed.), p. 203. W.A. Benjamin, New York, 1969 and unpublished results.

17. H. Hayashi, S. Iwata and S. Nagakura, *J. Chem. Phys.* <u>50</u>, 993 (1969).

18. H. Beens, J. de Jong and A. Weller, Colloque Ampere XV, Amsterdam 289 (1969).

19. P. Krebs, Ph.D. Thesis, University of Stuttgart (1973).

20. C.A. Parker and C.G. Hatchard, *Trans. Faraday Soc.* $\underline{59}$, 284 (1963).

21. H. Staerk, Ph.D. Thesis, Techn. University München (1969).

# THE PHOTOPHYSICS OF AROMATIC EXCIMERS

J.B. BIRKS

*The Schuster Laboratory, University of Manchester*

*Manchester, U.K.*

## EXCIMER INTERACTIONS

The original definitions of an excimer (1) and an exciplex (2) are applicable in fluid systems where the entities are free to associate and dissociate. The definitions require slight revision in rigid or intramolecular systems, where there are environmental or steric restraints on molecular motion. An excimer or exciplex is redefined as

"A molecular dimer or stoichiometric complex which is <u>associated</u> in an excited electronic state and which is <u>dissociative</u> (*i.e.* would dissociate in the absence of external restraints) in its ground electronic state".

The revised definitions include such entitites as the pyrene crystal dimer, the anthracene sandwich dimer in rigid solution, and intramolecular excimers, which are excluded by the original definitions.

The excimer association is characterised spectroscopically by its diffuse luminescence spectrum which is at lower energies than the molecular luminescence spectrum. In fluid systems the ground-state dissociation is shown by the absence of vibrational structure in the excimer luminescence spectrum, and by a molecular absorption spectrum. In rigid or intermolecular systems where the dissociation is inhibited, the excimer luminescence spectrum may exhibit some vibrational structure, and the molecular absorption spectrum may be perturbed by ground-state interaction.

A typical aromatic molecule has a singlet ground state, $S_o$, and a series of excited electronic singlet states, $S_p$, and triplet states, $T_q$, of zero-point energies $S_p^o$ and $T_q^o$, respectively. The magnitude of the interaction between an excited molecule and an unexcited molecule of the same species depends on the separation distance, $r$, and the relative orientation. Exciton resonance interaction splits each excited molecular singlet state, $S_p$, into a pair of singlet exciton resonance states of energies

$$^1(E_p)_{g,u} = S_p^o \mp \frac{m_p^2}{r^3} (\cos \alpha - 3 \cos \theta_1 \cos \theta_2) \qquad [1]$$

where $m_p$ is the $S_o$ - $S_p$ electric dipole transition moment, $\alpha$ is the angle between the transition dipoles in the two molecules, $\theta_1$ and $\theta_2$ are the angles between the dipoles and the line of centers, and the negative and positive signs refer to the even- and odd-parity states, $^1(E_p)_g$ and $^1(E_p)_u$, respectively.

If the two molecules are parallel ($\alpha = 0$, $\theta_1 = \theta_2 = \theta$), equation [1] simplifies to

$$^1(E_p)_{g,u} = S_p^o \mp \frac{m_p^2}{r^3} (1 - 3\cos^2 \theta) \qquad [2]$$

For a given value of $r$, $^1(E_p)_{g,u}$ has minimum and maximum values at $\theta = 90^o$, corresponding to a symmetrical sandwich configuration, of

$$^1(E_p)_{g,u} = S_p^o \mp \frac{m_p^2}{r^3} \qquad [3]$$

and maximum and minimum values at $\theta = 0^o$, corresponding to a parallel coplanar configuration, of

$$^1(E_p)_{g,u} = S_p^o \mp \frac{2m_p^2}{r^3} \qquad [4]$$

For spherical atoms or molecules, the lowest energy exciton resonance state is $^1(E_p)_u$ at $\theta = 0^o$ given by equation [4], and this applies to the excimers of the noble gases. For planar aromatic molecules, the closest distance of approach between molecular centers is much less in the symmetrical sandwich configuration ($\theta = 90^o$) than in the parallel coplanar configuration ($\theta = 0^o$), so that the lowest energy exciton resonance state is $^1(E_p)_g$ at $\theta = 90^o$ given by equation [3].

Charge-resonance states are due to coulombic interaction between positive ($^2M^+$) and negative ($^2M^-$) molecular ions. In the absence of orbital overlap between the two molecules, the four charge-resonance states are degenerate with a common energy

$$^{1,3}(R)_{g,u} = I - A - C(r) \qquad [5]$$

where $I$ is the molecular ionization potential, $A$ is the electron affinity, and $C(r)$ is the isotropic coulombic interaction potential,

$$C(r) = \frac{e^2}{\varepsilon r} \qquad [6]$$

where $e$ is the electronic charge and $\varepsilon$ is the dielectric constant of the solvent.

Configuration interaction of each exciton-resonance state, $^1(E)$, with the corresponding charge-resonance state, $^1(R)$,

yields a pair of mixed singlet excimer states, $^1(E \mp R)$, of
which the $^1(E - R)$ states have the lower energies. Figure 1
plots the calculated energies (3) of the $^1(E - R)$ states of
the symmetrical sandwich $(D_{6h})$ benzene dimer as a function of
the interplanar separation $r$. In benzene the lowest singlet
excimer state, $^1D^* = {}^1B_{1g} = {}^1(E_1 - R)_g$, corresponds to the

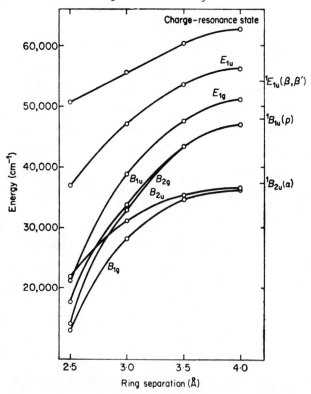

Fig. 1  *Singlet states of benzene excimer.*
*Theoretical energies of $D_{6h}$ sandwich dimer*
*as a function of interplanar separation r.*
*Low-energy excimer states of $^1B_{2u}$, $^1B_{1u}$ and*
*$^1E_{1u}$ molecular parentage [after Vala,*
*Hillier, Rice and Jortner (3)].*

lower exciton resonance state $^1(E_1)_g$ originating from
$S_1 = {}^1B_{2u}$ ($= {}^1L_b$) and stabilized by configuration interaction
with $^1(R)_g$.

$^1D^*$ does not usually originate from the $^1L_b$ molecular state, which has only a weak transition moment from $S_o$. In higher aromatic hydrocarbons in which $S_1 = ^1L_b$, the spacing of the molecular electronic energy levels is reduced, and $^1D^*$ originates from a higher molecular excited state, $S_p$, with a larger $S_o - S_p$ transition moment $m_p$. In naphthalene, $^1D^*$ originates from $S_2 = ^1L_a$, in 1,2-benzanthracene from $S_3 = ^1B_b$ and in pyrene from $S_4 = ^1B_a$ (4 - 6). In anthracene and the higher polyacenes, $^1D^*$ originates from $S_1 = ^1L_a$, which has a relatively strong transition moment from $S_o$. The influence of the resultant strong exciton resonance interaction on excimer and photodimer formation is discussed later.

Although the splitting of the triplet exciton resonance states

$$^3(E_q)_{g,u} = T^o_q \mp \frac{m^2_q}{r^3} \qquad [7]$$

is negligible in the aromatic hydrocarbons because $m_q \simeq 0$, configuration interaction with the triplet charge-resonance states, $^3(R)_{g,u}$, increases the splitting and yields mixed $^3(E \mp R)$ triplet excimer states, of which the lower $^3(E - R)$ states are associative. Figure 2 plots the calculated energies (7) of the $^3(E - R)$ states of the symmetrical sandwich $(D_{6h})$ benzene dimer as a function of the interplanar separation $r$. The lowest triplet excimer state, $^3D^* = ^3B_{2g} = ^3(E_1 - R)_g$, corresponds to the lower exciton resonance state, $^3(E_1)_g$, originating from $T_1 = ^3B_{1u}$ $(= ^3L_a)$ and stabilized by configuration interaction with $^3(R)_g$. The evidence for triplet excimers is discussed later.

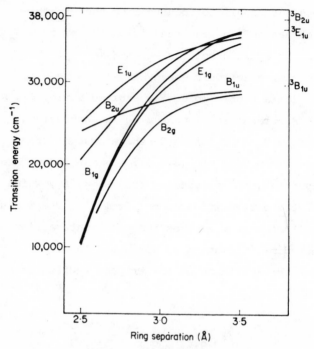

Fig. 2  *Triplet states of benzene excimer.*
*Theoretical energies of $D_{6h}$ sandwich dimer*
*as a function of interplanar separation r.*
*Low-energy excimer states of $^3B_{1u}$, $^3E_{1u}$ and*
*$^3B_{2u}$ molecular parentage [after Hillier,*
*Glass and Rice (7)].*

## THE EXCIMER POTENTIAL DIAGRAM

Figure 3 is a schematic potential energy diagram of a pair of parallel singlet-excited ($^1M^*$) and unexcited ($^1M$) molecules of the same species as a function of $r$. $V'(r)$ is the excimer interaction potential of the lowest singlet excimer state $^1D^*$. $R(r)$ and $R'(r)$ are the intermolecular repulsive potentials in the ground and excited states, respectively, and it is normally assumed that $R(r) = R'(r)$. The resultant singlet excimer energy is

$$D'(r) = V'(r) + R'(r) \qquad [8]$$

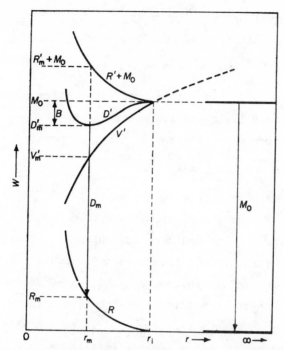

Fig. 3  Schematic potential energy (W) diagram of
parallel molecules $^1M^*$ and $^1M$ as a function
of intermolecular separation (r). R, R',
repulsive potentials in ground ($^1M$) and
excited ($^1M^*$) states; V', excimer interaction potential; D' (= V' + R'), resultant
excimer energy; $M_0$, molecular 0-0 transition; $D_m$, peak excimer transition; $R_m$,
$R'_m$, $V'_m$, $D'_m$, potentials at equilibrium excimer separation, $r_m$; B, excimer binding
energy.

$M_0$ (= $S_1^0$) is the $^1M^*$ energy corresponding to the 0-0 $S_1$ - $S_0$
fluorescence transition, and $D_m$ is the energy of the $^1D^*$
fluorescence maximum.  For excimer formation to occur, the
parameter

$$B(r) = M_0 - D'(r) = M_0 - V'(r) - R'(r)$$

$$= \Delta E_{DM}(r) - R'(r) \qquad [9]$$

must have a positive maximum of B, the $^1D^*$ binding energy, at

$r = r_m$, the equilibrium $^1D^*$ separation. $R_m$, $R_m' (= R_m)$, $V_m' (= D_m)$ and $D_m'$ are the potentials at $r = r_m$, so that

$$B = M_o - D_m - R_m \qquad [9a]$$

$\Delta E_{DM} (r) = M_o - V'(r)$ is the excimer associative potential, which must exceed the repulsive potential $R'(r)$ for excimer formation to occur. At $r = r_m$, $\Delta E_{DM} (r)$ becomes

$$\Delta E_{DM} = M_o. - D_m \qquad [9b]$$

the $^1M^* - ^1D^*$ spectroscopic energy gap.

There are two groups of aromatic hydrocarbons which do not form excimers or photodimers and which do not exhibit concentration quenching of fluorescence in fluid solutions. The first group includes molecules like 9,10-diphenylanthracene in which there is steric hindrance to the close approach of $^1M^*$ and $^1M$, so that $R'(r) > M_o - V'(r)$ at all values of $r$. The second group includes molecules like phenanthrene and chrysene in which the excimer interaction originating from a higher molecular state $S_p$ is too weak, i.e. $V'(r)$ is too large, so that $V'(r) > M_o - R'(r)$ at all values of $r$.

In a compound like 1,3-diphenylpropane which forms intramolecular excimers, the separation and orientation of the interacting phenyl groups are subject to the steric constraint of the linking alkane chain (8). This is equivalent to an increase in $R'(r)$, and also to an increase in $V'(r)$, due to the inhibition of a parallel configuration, so that the binding energy $B$ is less than that of the corresponding intermolecular excimer. In a rigid medium there may be external constraints on the molecular motion, equivalent to an increase in $R'(r)$ and a consequent decrease in $B$.

DIMERIC SPECIES

The types of dimeric species which can be formed from a pair of identical aromatic molecules may be classified as follows.

(i) A *dimer complex*, $^1(M.M)$, may be formed by inter-action between two unexcited molecules

$$^1M + {}^1M \rightleftharpoons {}^1(M.M) \qquad [10]$$

$^1(M.M)$ is normally also associated in its excited triplet, $^3(M.M)^*$, and singlet, $^1(M.M)^*$, states. The absorption, phosphorescence and fluorescence spectra of $^1(M.M)$ are more diffuse and at lower energies than those of the parent molecule.

(ii) A *triplet excimer*, $^3D^*$, may be formed by inter-action between a triplet-excited molecule, $^3M^*$, and an unexcited molecule

$$^3M^* + {}^1M \rightleftharpoons {}^3D^* \qquad [11]$$

or otherwise. $^3D^*$ has a dissociative ground state and an associated excited singlet state $^1D^*$. Its phosphorescence spectrum is diffuse and at lower energies than that of the parent molecule.

(iii) A *singlet excimer*, $^1D^*$, may be formed by inter-action of a singlet-excited molecule, $^1M^*$, and an unexcited molecule, $^1M$, as

$$^1M^* + {}^1M \rightleftharpoons {}^1D^* \qquad [12]$$

or otherwise. $^1D^*$ has a dissociative ground state, and it may undergo intersystem crossing to $^3D^*$, which is usually also dissociative. The $^1D^*$ fluorescence spectrum is diffuse and at lower energies than that of the parent molecule.

47

(iv) An *excited singlet excimer*, $^1D^{**}$, may be formed by interaction of a higher singlet-excited molecule, $^1M^{**}$, and an unexcited molecule, $^1M$

$$^1M^{**} + {}^1M \rightleftharpoons {}^1D^{**} \tag{13}$$

or otherwise. $^1D^{**}$ undergoes internal conversion to $^1D^*$ or dissociates.

(v) A *photodimer*, $^1M_2$, may be formed by the chemical interaction of a singlet-excited molecule, $^1M^*$, and an unexcited molecule, $^1M$,

$$^1M^* + {}^1M \longrightarrow {}^1M_2 \tag{14}$$

$^1M_2$ is stable in the ground state, but it normally dissociates in the excited singlet state $^1M_2^*$. Figure 4 shows the structure of

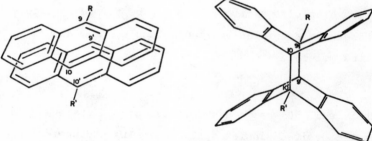

Fig. 4   *Structural diagrams of (a) the anthracene sandwich dimer (left) and (b) the anthracene photodimer, dianthracene (right) [after Tomlinson, Chandross, Fork, Pryde and Lamola (9)].*

(a) a sandwich pair of anthracene molecules, and

(b) the photodimer, dianthracene, formed therefrom (9).

The molecular structure of the photodimer (b) indicates that the symmetrical sandwich configuration (a) is that of the interacting molecules immediately prior to the reaction [14]. The $^1M_2$ absorption spectrum is at much shorter wavelengths than that of $^1M$ and it is unrelated thereto.

(vi) A *sandwich dimer*, $(M \parallel M)$, may be formed by the
photolysis of a photodimer, $^1M_2$, in a rigid matrix (9 - 11),
reversing process [14]

$$^1M_2 + h\nu \longrightarrow {}^1M_2^* \longrightarrow {}^1(M \parallel M)^* \longrightarrow (M \parallel M) \qquad [15]$$

and yielding a pair of adjacent molecules in a sandwich
configuration (figure 4a) which emit excimer ($^1D^*$) fluorescence. $(M \parallel M)$ configurations also occur naturally in
type B (*e.g.* pyrene and perylene) crystals.

(vii) A *nonsandwich dimer*, $(M \times M)$, corresponds to a
pair of identical molecules which are not in a sandwich or
displaced sandwich configuration, but which are sufficiently
close or suitably oriented to interact in their excited and/
or ground states.

## EXCIMER STERIC CONFIGURATIONS

Excimer ($^1D^*$) fluorescence may occur from various
$(M \parallel M)$ and $(M \times M)$ configurations depending on the conditions.

(i) In an optically excited fluid solution singlet excimers $^1D^*$ are in thermodynamic equilibrium with the dissociated excited molecular system ($^1M^* + {}^1M$), and both $^1D^*$ and
$M^*$ fluorescence is observed. The excimer energies lie
between the zero-point energies of $^1D^*$ ($= D_m'$) and $^1M^*$ ($= M_o$),
covering an energy range equal to the excimer binding energy
. This range includes all possible excimer configurations
for which the excimer energy

$$D' = V' + R' < M_o \qquad [16]$$

In the absence of steric hindrance the equilibrium excimer
configuration corresponding to the excimer fluorescence maximum $D_m$ is a $^1(M \parallel M)^*$ symmetrical sandwich structure, but

the more energetic excimer configurations include various other $^1(M \parallel M)^*$ and $^1(M \times M)^*$ structures.

(ii) The situation is similar in an excited aromatic liquid, such as benzene or toluene (12), at normal temperature where the rate of excimer formation and dissociation greatly exceeds the excitation decay rate. There is a rapid interchar between the dissociated ($^1M^* + {}^1M$) configurations, responsible for the molecular fluorescence, and the associated $^1(M \parallel M)^*$ and $^1(M \times M)^*$ configurations, responsible for the excimer fluorescence, due to thermal fluctuations in the relative positions of the excited molecules, $^1M^*$, and their unexcited neighbours, $^1M$.

(iii) In a pyrene crystal, which has a type B lattice, the molecules are arranged in $(M \parallel M)$ sandwich pairs with an intermolecular spacing of $r_o = 3.53$ Å. Following $^1M^*$ excitation, the excimer interaction reduces the spacing to $r_m = 3.34$ to form the equilibrium $^1(M \parallel M)^*$ excimer configuration (6) a shown in figure 5. The intermediate excimer states with energ between $^1D^*$ and $^1M^*$ are vibrationally excited states of the $^1(M \parallel M)^*$ configuration.

(iv) In an anthracene crystal, which has a type A latti there are two types of nearest-neighbour dimers: $(M \parallel M)$, corresponding to translationally equivalent molecules; and $(M \times M)$, corresponding to translationally inequivalent molecul The former are responsible for the exciton shift, and the latt for the Davydov splitting, of the crystal absorption spectrum In neither dimer configuration in the crystal lattice are the molecules sufficiently close or suitably oriented for the excimer interaction responsible for these effects to exceed the lattice forces, so that in a perfect crystal the fluorescence is molecular and no excimers are formed. If the crystal conta defects or vacancies, these provide sites where adjacent molecules have different orientations and more freedom of movement than

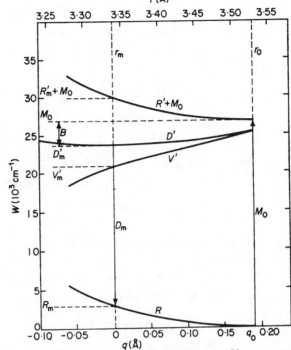

Fig. 5  *Experimental potential energy diagram of pyrene crystal dimer and excimer. Energy (W) against intermolecular separation (r) and displacement (q) from excimer equilibrium $(r_m)$. Notation as in figure 3 [after Birks and Kazzaz (6)].*

in the perfect lattice.  $^1(M \parallel M)^*$  and  $^1(M \times M)^*$  excimers may be formed at these sites, which function as traps for  $^1M^*$  excitons, so that an excimer defect emission band appears in the crystal fluorescence spectrum (13).  Photodimers  $(^1M_2)$  are also formed at such sites;  anthracene crystals photodimerize preferentially at dislocations (14, 15).

(v)  Intramolecular excimers (8) in fluid solution behave in a similar manner to intermolecular excimers, except that the molecular chain linking the two interacting chromophores imposes a steric restraint on their possible relative positions. Chandross and Dempster (16) studied the fluorescence of various dinaphthylalkanes, but they only observed strong intramolecular excimer fluorescence in the two 1,3-dinaphthylpropanes, from which they concluded that the optimum excimer configuration is $^1(M \parallel M)^*$.

(vi) The excimer fluorescence of aromatic polymers (8) is due to intermolecular and/or intramolecular $(M \times M)$ and $(M \parallel M)$ dimers, suitably oriented and with sufficient freedom of motion to form excimers. These sites act as traps for $^1M^*$ excitons which migrate along and between the polymer chains.

## MODES OF EXCIMER FORMATION

Excimers can be formed by processes other than the direct interaction of excited and unexcited molecules, reactions [11] - [13].

### DIMER CATION NEUTRALIZATION

The interaction of a molecular cation, $^2M^+$, with a neutral molecule, $^1M$, yields a metastable dimer cation, $^2D^+$, as

$$^2M^+ + {}^1M \longrightarrow {}^2D^+ \qquad [17]$$

The subsequent neutralization of $^2D^+$

$$^2D^+ + {}^2e^- \longrightarrow {}^{1,3}D^{**} \qquad [18]$$

yields excited singlet and triplet excimers, which internally convert to singlet and triplet excimers

$$^{1,3}D^{**} \longrightarrow {}^{1,3}D^* \qquad [19]$$

This sequence of processes has been used to explain the predominant excimer fluorescence of liquid alkylbenzenes excited by an intense electron beam (17).

## CATION-ANION ASSOCIATION

The association of molecular cations, $^2M^+$, and anions, $^2M^-$, represents an alternative mode of excimer formation

$$^2M^+ + \,^2M^- \Big\langle \begin{array}{ll} ^{1,3}D^* & \text{[20a]} \\[1em] ^{1,3}M^* + \,^1M & \text{[20b]} \end{array}$$

The interpretation of electroluminescence studies based on these processes has been discussed previously (4).

## TRIPLET-TRIPLET INTERACTION

The P-type delayed molecular and excimer fluorescence of aromatic molecules in fluid solutions originates from the following processes (4):

$$^3M^* + \,^3M^* \Big\langle \begin{array}{ll} ^1M^* + \,^1M & \text{[21a]} \\[1em] ^1D^* & \text{[21b]} \end{array}$$

The ratio $\phi_{FD}^d/\phi_{FM}^d$ of the delayed $^1D^*$ and $^1M^*$ fluorescence yields exceeds the ratio $\phi_{FD}/\phi_{FM}$ of the prompt $^1D^*$ and $^1M^*$ fluorescence yields, showing that reactions [21a] and [21b] are parallel, and not consecutive, processes (4). $\phi_{FM}^d$ and $\phi_{FD}^d$ depend on

(i) $k_{MTT}$ and $k_{DTT}$, the rate parameters of processes [21a] and [21b], respectively;

(ii) $k_{DM}[^1M]$ and $k_{MD}$, the rates of $^1D^*$ formation from, and dissociation to, $^1M^* + \,^1M$, respectively;

(iii) $k_M$, $k_D$ and $k_T$, the $^1M^*$, $^1D^*$ and $^3M^*$ unimolecular decay rates, respectively; and

(iv) the excitation light intensity.

In general (4)

$$\phi_{FM}^{d} = a\{k_D \ k_{MTT} + k_{MD} \ (k_{MTT} + k_{DTT})\} \qquad [22]$$

$$\phi_{FD}^{d} = b\{k_M \ k_{DTT} + k_{DM}[^1M] \ (k_{MTT} + k_{DTT})\} \qquad [23]$$

Since $^3M^*$ is paramagnetic, the rates of processes [21a] and [21b] may be influenced by an applied magnetic field $H$. Wyrsch and Labhart (18) have studied the effect of $H$ on the delayed emission intensities of dilute ethanolic solutions of 1,2-benzanthracene from -70 to -160°C. Typical results are shown in figure 6. The molecular phosphorescence intensity $\phi_{PT}$ is independent of $H$; the delayed $^1D^*$ fluorescence

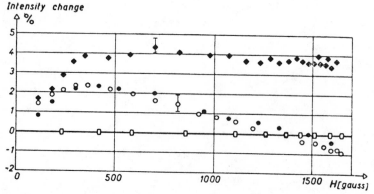

Fig. 6 *1,2-Benzanthracene in dilute ethanol solution. Relative changes of intensities of phosphorescence ($\phi_{PT}$), delayed molecular fluorescence ($\phi_{FM}^{d}$) and delayed excimer fluorescence ($\phi_{FD}^{d}$) as a function of magnetic field strength H.[after Wyrsch and Labhart (18)].*

□ $\phi_{PT}$ *at* -133°C       ○ $\phi_{FM}^{d}$ *at* -133°C

● $\phi_{FM}^{d}$ *at* -107°C       ◆ $\phi_{FD}^{d}$ *at* -107°C

intensity $\phi_{FD}^{d}$ increases by about 4% to a limiting value at $H \geq 0.4$ kG; and the delayed $^1M^*$ fluorescence intensity $\phi_{FM}^{d}$ increases by about 2% to a maximum at $H \sim 0.4$ kG and then decreases monotonically to below its zero-field value at $H > 1.65$ kG. For dilute ethanolic solutions of

1,2-benzanthracene (18) below $-70^{\circ}C$, $k_M$, $k_D \gg k_{DM}[^1M]$, $k_{MD}$
so that equations [22] and [23] reduce to

$$\phi^d_{FM} = a\, k_D\, k_{MTT} \qquad\qquad [24]$$

$$\phi^d_{FD} = b\, k_M\, k_{DTT} \qquad\qquad [25]$$

The observed magnetic field dependence of $\phi^d_{FM}$ and $\phi^d_{FD}$
(figure 6) corresponds to that of $k_{MTT}$ and $k_{DTT}$, respectively,
since the other parameters in equations [24] and [25] are
independent of $H$.

Process [21a], with rate parameter $k_{MTT}$, is due to a
medium range ($\sim 6 - 12$ $\overset{o}{A}$) electron-exchange interaction. It
is responsible for the delayed molecular fluorescence of
concentrated rigid solutions and of type A aromatic crystals
(4). The magnetic field dependence $\phi^d_{FM}(H)$ of crystal anthracene (19) is similar to that of dilute low-temperature
1,2-benzanthracene solutions (figure 6).

Process [21b], with rate parameter $k_{DTT}$, is due to a short
range collisional interaction. The parameter

$$\alpha = \frac{k_{DTT}}{k_{MTT}} \qquad\qquad [26]$$

depends on the solution viscosity $\eta$ or, for a given solvent,
on the temperature. At low viscosities $\alpha$ ($= 0.9$ for
1,2-benzanthracene, 2.0 for pyrene) is independent of $\eta$, but
with an increase in viscosity, $\alpha$ decreases towards zero. The
viscosity dependence of $\alpha$ is due to the difference in interaction range of the two processes. The short range process,
$k_{DTT}$, is diffusion-controlled, while the medium range process,
$k_{MTT}$, may be considered to consist of a diffusion-controlled
component, $k'_{MTT}$, and a diffusion-independent component, $k^o_{MTT}$,
so that

$$\alpha = \frac{(\alpha)_{\infty}}{1 + k_{MTT}^{o}/k_{MTT}'} \qquad [26a]$$

where $(\alpha)_{\infty} = k_{DTT}/k_{MTT}'$ is the value of $\alpha$ at low $\eta$ (4, 20).

In dilute low viscosity solutions of anthracene, $\phi_{FM}^{d}$ decreases monotonically with an increase in $H$ (21), in contrast to the behavior in high viscosity solutions of 1,2-benzanthracene (figure 6). In anthracene solutions, $\phi_{FD}^{d}$ (= 0) and $k_{DTT}$ are replaced by $\phi_{A}^{d}$ and $k_{ATT}$, the quantum yield and rate parameter of $^{1}M_{2}$ formation, respectively. If the magnetic field dependence of $k_{ATT}$ is similar to that of $k_{DTT}$ (figure 6), the form of $\phi_{FM}^{d}(H)$ for low-viscosity anthracene solutions may be explained by competition between $k_{MTT}$ and $k_{ATT}$.

To determine $k_{MTT}(H)$, $\phi_{FM}^{d}(H)$ should preferably be observed under solution conditions where the processes $k_{DTT}$ and $k_{ATT}$, competitive with $k_{MTT}$, are negligible. For molecules which form excimers or dimers, this involves very viscous or rigid solutions. For phenanthrene and 9,10-diphenylanthracene, where $k_{DTT} = k_{ATT} = 0$, there are no such restrictions on the solution viscosity.

PHOTODIMERS AND EXCIMERS

In the majority of the *cata*-condensed aromatic hydrocarbons (except anthracene and the higher polyacenes) $S_{1} = {}^{1}L_{b}$, which has only a weak $S_{o} - S_{1}$ transition moment. In benzene, charge-resonance configuration interaction is sufficient to stabilize the $^{1}(E_{1})_{g}$ state originating from $S_{1}$ (figure 1), but in higher hydrocarbons $^{1}D^{*}$ corresponds to a $^{1}(E_{p})_{g}$ state originating from a higher molecular state $S_{p}$ with a larger transition moment $m_{p}$, as discussed in the first section. The

$^1(E_p)_g$ energy is a minimum in the symmetrical $^1(M \parallel M)^*$ configuration, but it is only associative, $(\Delta E_{DM} (r) > 0)$, when

$$r \leq r_D = \left( \frac{m_p^2}{s_p^o - s_1^o} \right)^{1/3} \qquad [27]$$

where $r_D$ is the effective range of the excimer interaction (5).

When $S_1 = {}^1L_a$, which has a relatively strong transition moment $m_1$, $^1(E_1)_g < s_1^o$ for all values of $r$. The resultant attractive potential is a maximum in the symmetrical $^1(M \parallel M)^*$ configuration, equation [3], and it is given by

$$\Delta E_{DM} (r) = s_1^o - {}^1(E_1)_g = \frac{m_1^2}{r^3} \qquad [28]$$

Charge-resonance configuration interaction further increases the attractive potential. In photoexcited fluid solutions of anthracene, tetracene and pentacene for which $S_1 = {}^1L_a$, the attractive potential, equation [26], aligns $^1M^*$ and $^1M$ in a close symmetrical $^1(M \parallel M)^*$ configuration (figure 4a). Here they interact chemically to form two covalent bonds linking the *meso*-positions of the two molecules, the regions of high electron density in the $^1L_a$ state, and thus form the photo-dimer $^1M_2$ (figure 4b). The four valence electrons in each of the four *meso* carbon atoms change from an unsaturated ($sp^2$ hybrid, $\pi^*$) electron configuration to a saturated ($sp^3$ hybrid) configuration.

Ferguson and Mau [22] have made an elegant study of the photophysics of the anthracene sandwich dimers, $(M \parallel M)$, formed by the photolysis of dianthracene molecules, $^1M_2^*$, in a dianthracene crystal by process [15]. Photoexcitation of $(M \parallel M)$ leads to the following processes

$$^1_{(M \,||\, M)}{}^* \xrightarrow{\phantom{xxxx}} {}^1_{D}{}^* \xrightarrow{\;\;k_{AD}\;\;} {}^1_{M_2}$$

$$\Big\downarrow k_{FD} \qquad\qquad [29]$$

$$^1_{(M \,||\, M)} + h\nu_D$$

The quantum yields of $^1_D{}^*$ fluorescence, $\phi_{FD}$ $(= k_{FD}/k_D)$, and of $^1_{M_2}$ formation, $\phi_{AD}$ $(= k_{AD}/k_D)$, are observed to sum to unity at all temperatures from 4 to 300 K. At low temperatures, $\phi_{FD} = 1.0$ and $\phi_{AD} = 0$; at room temperature $\phi_{FD}$ tends to zero and $\phi_{AD}$ tends to unity. In the relation

$$k_{AD} = k'_{AD} \exp\left(- W_{AD}/kT\right) \qquad [30]$$

$k'_{AD} = 7.6 \times 10^9$ s$^{-1}$ and $W_{AD} = 590$ cm$^{-1}$. The results show that excimer fluorescence and photodimerization are the only significant processes competing for the $^1_D{}^*$ excitation energy, and they confirm that the excimer is the intermediate state in photodimer formation.

In fluid solution, the 9-substituted anthracenes exhibit both excimer fluorescence and photodimerization. Birks and Aladekomo (23) proposed that the excimer fluorescence of 9-methylanthracene in solution occurs from the *cis* (head-to-head) excimer, $^1_{D_c}{}^*$, and that the photodimer $^1_{M_2}$ is formed from the *trans* (head-to-tail) excimer, $^1_{D_t}{}^*$. Other solution and crystal studies support this hypothesis. All the 9-substituted anthracenes studied in room temperature solution yield photodimers with *trans* configurations (24 - 28). 9-Cyanoanthracene (9-CNA) and similar compounds, crystallize in a *cis* (M || M) configuration, and the crystals exhibit strong *cis*-excimer ($^1D_c{}^*$) fluorescence. Only *trans*-photo-dimers are formed within the crystal, but these are produced at defects where the molecular arrangement is disordered (29 - 32). The *trans* sandwich dimer of 9-CNA, formed by photolysis of the *trans*-photodimer in rigid solution, exhibits only weak excimer ($^1D_t{}^*$) fluorescence, and it reverts rapidly to the *trans*-photodimer on exposure to light (33).

Anthracene and its 9-methyl, 9,10-dimethyl and 9,10-diphenyl derivatives each have $S_1 = {}^1L_a$ with similar excimer attractive potentials, $m_1^2/r^3$, in the symmetric (M $\parallel$ M) configuration. The difference in their ability to form photodimers or excimers in fluid solutions is due to the steric factors which determine the minimum intermolecular spacing $r_{min}$. $r_{min}$ is least for the sandwich dimers of anthracene and of 9-methylanthracene in the *trans* configuration, and these compounds photodimerize; $r_{min}$ is larger for the sandwich dimers of 9-methylanthracene in the *cis* configuration and of 9,10-dimethylanthracene, and these compounds form excimers, but not photodimers. $r_{min}$ is largest for the sandwich dimers of 9,10-diphenylanthracene (the phenyl groups are at $60^{\circ}$ to the anthracene plane), and this compound does not form excimers or photodimers. It does, however, give P-type delayed fluorescence by process [21b], demonstrating that the $^3M^* - {}^3M^*$ interaction [24a] is of longer range than the excimer interaction (34).

PHOTOPHYSICAL PROCESSES IN EXCIMERS

The reaction scheme and corresponding rate parameters of an excited molecular and excimer system are shown in figure 7. $^1M^*$ decays by fluorescence ($k_{FM}$), internal conversion ($k_{GM}$), intersystem crossing to $^3M^*$ ($k_{TM}$) and $^1D^*$ formation

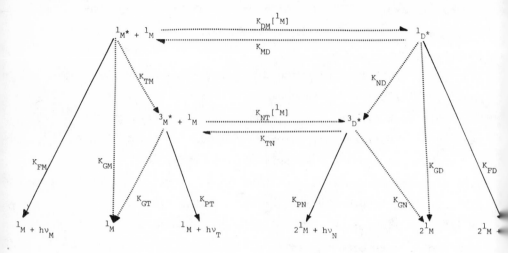

*Fig. 7   Molecular and excimer rate processes and parameters.*

$(k_{DM}[^1M])$.   $^1D^*$ decays by fluorescence $(k_{FD})$, dissociative internal conversion $(k_{GD})$, intersystem crossing to $^3D^*$ $(k_{ND})$ and dissociation to $^1M^* + ^1M$ $(k_{MD})$.   $^3M^*$ decays by phosphorescence $(k_{PT})$, intersystem crossing $(k_{GT})$ and $^3D^*$ formation $(k_{NT}[^1M])$.   $^3D^*$ decays by phosphorescence $(k_{PN})$, dissociative intersystem crossing $(k_{GN})$ and dissociation to $^3M^* + ^1M$ $(k_{TN})$.

In solution systems, $k_M$ $(= k_{FM} + k_{GM} + k_{TM})$, $k_D$ $(= k_{FD} + k_{GD} + k_{ND})$, $k_{DM}$ and $k_{MD}$ are evaluated from the concentration dependence of the $^1M^*$ and $^1D^*$ fluorescence response functions (4).   The parameters $k_{FM}$, $k_{IM}$ $(= k_{GM} + k_{TM})$ $k_{FD}$ and $k_{ID}$ $(= k_{GD} + k_{ND})$ are then obtained from the concentration dependence of the $^1M^*$ and $^1D^*$ fluorescence quantum yields. Observations of the concentration dependence of the triplet $(^3M^*)$ quantum yield (35, 36) enable $k_{GM}$, $k_{TM}$, $k_{GD}$ and $k_{ND}$ to be evaluated (37), assuming that $^3D^*$ dissociates into $^3M^* + ^1M$.   The temperature dependence of the molar equilibrium constant $(k_{DM}/k_{MD})$ gives the enthalpy and entropy of $^1D^*$ formation.   The temperature dependence of the various rate parameters gives their frequency factors and activation energies. In general, the $^1D^*$ internal quenching rate parameter

$$k_{ID} = k_{GD} + k_{ND} = k^o_{ID} + k'_{ID} \exp\left(- W_{ID}/kT\right) \qquad [31]$$

but the assignment of the temperature-independent and temperature-dependent components to $k_{GD}$ and $k_{ND}$ depends on the conditions.

For the pyrene crystal excimer $k_{FD} = 5.5 \times 10^6$ s$^{-1}$, $k^o_{ID} \simeq 0$, $k'_{ID} = 4.7 \times 10^6$ s$^{-1}$ and $W_{ID} = 0.066$ eV (38). The magnitude of $k'_{ID}$ is consistent with intersystem crossing, and the $^1D^*$ and $(^3M^* + ^1M)$ potential curves of the pyrene crystal (figure 5) cross at about 0.1 eV ($\sim W_{ID}$) above the $^1D^*$ potential minimum. Only three luminescences are observed from pure pyrene crystals (39):

prompt $^1D^*$ fluorescence $(k_{FD})$;
$^3M^*$ phosphorescence $(k_{PT})$; and
delayed $^1D^*$ fluorescence due to $^3M^*$ - $^3M^*$ interaction, process [21b].

An emission previously attributed to $^3D^*$ phosphorescence (40 - 42) is due to impurities. It is concluded that intersystem crossing is the predominant radiationless process in the pyrene crystal singlet excimer $^1D^*$, $(k_{ID} \simeq k_{ND};\ k_{GD} \simeq 0)$, and that the triplet excimer $^3D^*$ dissociates into $^3M^* + ^1M$.

For pyrene solutions in ethanol and $n$-hexane, $k_{ND}$ and $k_{GD}$ have been evaluated (37) from experimental data. In each solvent, $k_{ND}$ ($= k^o_{ID}$) is found to be temperature independent, while $k_{GD}$ ($= k'_{ID} \exp\left(- W_{ID}/kT\right)$) is temperature dependent. In various solvents, $W_{ID}$ is found to correspond to the activation energy for viscous motion. The thermally-activated process, $k_{GD}$, is therefore attributed to molecular motion which distorts the excimer from a $^1(M \parallel M)^*$ to a $^1(M \times M)^*$

configuration, in which increased vibrational interaction between the molecules at their closest distance of approach leads to internal conversion to the dissociated ground state. Such a configuration change cannot occur in the pyrene crystal, where $k_{GD} \simeq 0$, and it also appears to be inhibited in fluid solution when the viscosity exceeds 4.7 cP (37).

For 1-methylnaphthalene in ethanol solution, Cundall and Pereira (43) determined values of $k_{ID}^O = 5 \times 10^6$ s$^{-1}$, $k_{ID}' = 3.2 \times 10^{11}$ s$^{-1}$ and $W_{ID} = 0.29$ eV. From the magnitudes of $k_{ID}^O$ and $k_{ID}'$, they attributed $k_{ID}^O$ to the spin-forbidden intersystem crossing process, $k_{ND}$, and $k_{ID}'$ to the spin-allowed internal conversion process $k_{GL}$, as in the pyrene solution excimer (37).

## TRIPLET EXCIMERS

The condition for singlet excimer formation is that

$$B(r) = \Delta E_{DM}(r) - R'(r) \qquad [9]$$

should have a positive maximum $B$ at $r = r_m$, at which

$$\Delta E_{DM} = M_o - D_m \qquad [9b]$$

is the $^1M^*$ - $^1D^*$ spectroscopic energy gap. The corresponding condition for triplet excimer formation is that

$$B_N(r) = \Delta E_{NT}(r) - R'(r) \qquad [32]$$

should have a positive maximum $B_N$ at $r = r_N$, at which

$$\Delta E_{NT} = T_1^O - N_m \qquad [32b]$$

is the $^3M^*$ - $^3D^*$ spectroscopic energy gap. $\Delta E_{NT}(r) = T_1^O - V_N'(r)$ is the $^3D^*$ associative potential, $V_N'(r)$ is the $^3D^*$ interaction potential and $N_m$ is the energy of the $^3D^*$ phosphorescence maximum.

In general, $^3D^*$, the lowest triplet excimer state, is the $^3(E_1)_g$ exciton resonance state originating from $T_1$ as stabilized by configuration mixing with the $^3(R)_g$ charge-resonance state. In the absence of such mixing the $^3(E_1)_{g,u}$ splitting is negligible in the aromatic hydrocarbons, equation [7]. Although configuration interaction increases the splitting (figure 2), the effect is less than for the singlet excimer states (figure 1) because of the larger energy gap $\Delta = {}^3(R)_g - T_1^o$ (the configuration interaction is proportional to $\Delta^{-2}$). It is concluded that

$$\Delta E_{NT} \ (r) \ll \Delta E_{DM} \ (r) \qquad [33]$$

and that consequently

$$B_N \ll B \qquad [33a]$$

$$\Delta E_{NT} \ll \Delta E_{DM} \qquad [33b]$$

$$r_N > r_m \qquad [33c]$$

These criteria may be used to assess the spectroscopic evidence for triplet excimers.

For benzene $D_m = 31,300 \ cm^{-1}$, $\Delta E_{DM} = 5,700 \ cm^{-1}$ (12) and $B = 2,800 \ cm^{-1}$ (44), corresponding to $R'_m = 2,900 \ cm^{-1}$. Comparison of $D_m$ with the calculated energy of the $^1B_{1g}$ excimer state (figure 1) gives $r_m = 3.3 \ \overset{o}{A}$. The calculated $^3B_{2g}$ excimer energy (figure 2) gives $\Delta E_{NT} \ (r_m) = 1,900 \ cm^{-1} < R'(r_m) = 2,900 \ cm^{-1}$. Extrapolation of $R'(r)$ suggests that it may exceed $\Delta E_{NT}(r)$ at all values of $r$, in which case benzene does not form triplet excimers. If it does so, it is estimated that $r_N > 3.5 \ \overset{o}{A}$, $\Delta E_{NT} < 1000 \ cm^{-1}$ and $B_N < 200 \ cm^{-1}$.

The 2,2'-paracyclophane molecule consists of two benzene rings joined by $(CH_2)_2$ chains at their para positions to form a slightly distorted (M ‖ M) configuration, with the separation distance $r$ varying from 2.8 to 3.09 $\overset{o}{A}$ (8). The fluorescence spectrum and the corresponding absorption spectrum are broad bands, with vibrational fine structure, due to the intramolecular singlet "dimer". The phosphorescence spectrum is a similar broad band with a maximum at $\sim 21,000 \text{ cm}^{-1}$, which agrees satisfactorily with the calculated benzene $^3B_{2g}$ excimer energy at $r = 3.0 \text{ Å}$ (7) (figure 2). The high value of $\Delta E_{NT} \sim 8,500 \text{ cm}^{-1}$ for the triplet "dimer" is due to the short intermolecular links.

In 4,4'-paracyclophane the two benzene rings are joined by $(CH_2)_4$ links and are separated by $r = 3.73 \text{ Å}$ (8). The absorption spectrum is structured and it shows negligible ground-state interaction. The fluorescence spectrum includes an intense structureless band ($D_m = 29,400 \text{ cm}^{-1}$) due to intramolecular singlet excimers produced by the closer approach of the benzene rings towards $r_m \simeq 3.3 \text{ Å}$. In contrast, the phosphorescence spectrum is structured and it has a bathochromic shift of only $500 \text{ cm}^{-1}$ relative to ethylbenzene (7). Despite the favorable (M $\parallel$ M) configuration, the negligible ground-state interaction and the formation of singlet excimers, no intramolecular excimer phosphorescence is observed.

A structureless emission band at $\sim 22,000 \text{ cm}^{-1}$ observed in the spectrum of solid benzene excited by 1 MeV electrons has been assigned to benzene excimer phosphorescence (45). The assignment is considered to be incorrect, since it would correspond to $\Delta E_{NT} \sim 6,000 \text{ cm}^{-1} \sim \Delta E_{DM}$, which is inconsistent with equation [32b]. The assignments of the structureless emission bands at $19,000 - 20,000 \text{ cm}^{-1}$ observed in the spectra of six liquid alkylbenzenes excited by an intense electron beam to triplet excimer emission (46) are also considered incorrect. These assignments would correspond to $\Delta E_{NT} \sim 8,000 - 9,000 \text{ cm}^{-1}$, which exceeds the values of $\Delta E_{DM} = 3,500 - 4,800 \text{ cm}^{-1}$ observed for the alkylbenzenes (4), in contradiction to equation [32b].

The *syn*- and *anti*- 2,2'-paracyclonaphthanes are the ana-
logues of 2,2'-paracyclophane in which two naphthalene molecules are joined by dimethylene groups at their $\alpha$-positions
in symmetric and displaced sandwich configurations, respectively, with a separation of $r \simeq 3$ Å (47). The *syn*- and
*anti*- isomers have structureless fluorescence spectra with
$D_m = 21,650$ cm$^{-1}$, $\Delta E_{DM} = 9,250$ cm$^{-1}$, and $D_m = 24,900$ cm$^{-1}$,
$\Delta E_{DM} = 6,000$ cm$^{-1}$, respectively. The *anti*-isomer fluorescence
spectrum agrees with that of the 1,4-dimethylnaphthalene excimer, but this is fortuitous since the latter has a symmetric
(M || M) configuration and a wider separation $r_m \simeq 3.3$ Å.
The phosphorescence spectra of the *syn*- and *anti*- isomers are
structured with bathochromic shifts of 2,700 cm$^{-1}$ and
1,700 cm$^{-1}$, respectively, relative to 1,4-dimethylnaphthalene.
Comparison of the structured phosphorescence spectra with
the structureless "dimer" phosphorescence spectrum of
2,2'-paracyclophane (7) indicates that naphthalene has less
tendency to form triplet excimers than benzene.

This conclusion is confirmed by the experiments of
Chandross and Dempster (48) who prepared the intramolecular
photodimer of 1,3-di-$\alpha$-naphthylpropane, and then photolysed
it in solid solution at 77 K to produce the intramolecular
sandwich dimer. The (M || M) dimer yields an excimer
fluorescence spectrum similar to that of naphthalene, and the
absence of molecular fluorescence shows that all the naphthalene groups are arranged in sandwich pairs. In contrast, the
phosphorescence spectrum is structured with only a small
bathochromic shift of 250 cm$^{-1}$, relative to the phosphorescence spectrum of the same specimen after destruction of the
sandwich structure by thawing and subsequent recooling. It
is concluded that naphthalene does not form triplet excimers,

unless these have a quite different configuration from the $^1(M \parallel M)^*$ singlet excimer, which is unlikely. A structureless emission band at $\sim 18,600$ cm$^{-1}$ observed in the spectrum of naphthalene in ethanol solution at 160 K has been attributed to excimer phosphorescence (49, 50), corresponding to $\Delta E_{NT} \simeq 2,6000$ cm$^{-1}$, compared with $\Delta E_{DM} = 6,500$ cm$^{-1}$ (4). A similar emission band observed in the spectra of phenanthrene solutions and originally attributed to excimer phosphorescence (49) was subsequently found to be due to an impurity (50), and the naphthalene emission may be due to a similar cause.

The pyrene crystal structure with its $(M \parallel M)$ configuration and negligible ground-state interaction provides ideal conditions for the observation of any excimer phosphorescence. Although emissions attributed to this cause have been reported (40 - 42), they have since been shown to be due to impurities (39). A pure pyrene crystal emits only excimer $(^1D^*)$ fluorescence and molecular $(^3M^*)$ phosphorescence, showing that the triplet excimer is not associated at $r = 3.53$ Å. The cumulative spectroscopic evidence indicates that benzene, naphthalene, pyrene and other aromatic hydrocarbons do not emit excimer phosphorescence.

The molecular triplet $(^3M^*)$ lifetimes $\tau_T$ of $o$-xylene (51), naphthalene, anthracene, phenanthrene and pyrene (50, 52) in fluid solution are observed to decrease with an increase in molar concentration, $[^1M]$.

$$1/\tau_T = k_T = (k_T)_o + k_{QT}[^1M] \qquad [34]$$

where $(k_T)_o = k_{PT} + k_{GT}$ is the $^3M^*$ decay rate at infinite dilution, and $k_{QT}$ is the self-quenching rate parameter. The triplet self-quenching has been attributed (50 - 52) to the formation, $k_{NT}[^1M]$, dissociation, $k_{TN}$ and decay,

$k_N$ (= $k_{PN}$ + $k_{GN}$) of triplet excimers as follows

$$^3M^* + {}^1M \underset{k_{TN}}{\overset{k_{NT}[^1M]}{\rightleftharpoons}} {}^3D^* \qquad [35]$$

$$\downarrow k_N$$

$$^1M + {}^1M$$

On this model

$$k_{QT} = \frac{k_N k_{NT}}{k_{TN} + k_N} = p\, k_{NT} \qquad [36]$$

where $k_{NT}$ is the diffusion-controlled rate parameter and $p$ is the quenching probability per encounter. Values of $p$ from $5 \times 10^{-5}$ to $3 \times 10^{-4}$ are observed in 20° C solutions (50, 52). Since $k_{TN} = k_N/p \gg k_N$, the results indicate a low binding energy, $B_N$, for the triplet excimers of the aromatic hydrocarbons.

The triplet self-quenching of phenanthrene in solution (50, 52) is unusual, since phenanthrene does not form singlet excimers or exhibit fluorescence self-quenching. If the phenanthrene contained a mole fraction $f$ of its usual impurity, anthracene ($^1A$), quenching would then occur by diffusion-controlled triplet-triplet energy transfer

$$^3M^* + {}^1A \xrightarrow{k_{AT}[^1A]} {}^1M + {}^3A^* \qquad [37]$$

at a rate

$$k_{AT}[^1A] = f\, k_{AT}[^1M] \qquad [38]$$

Since $k_{AT} \simeq k_{NT}$, $f \simeq p = 10^{-4}$ for phenanthrene (50), which is a reasonable impurity content. For naphthalene, anthracene and pyrene (50, 52), the values of $p$ of $5 \times 10^{-5}$ to $3 \times 10^{-4}$ are of similar magnitude to the typical impurity content $f$,

67

so that process [37] needs to be considered as a possible alternative to [35] to account for the "triplet self-quenching" in these compounds.

The halobenzenes do emit excimer phosphorescence. Such emission has been observed from crystals of 1,4-dichloroben-zene, 1,3,5-trichlorobenzene, 1,2,4,5-tetrachlorobenzene, 1,4-dibromobenzene and 1,3,5-tribromobenzene (53, 54) and from concentrated rigid solutions of chlorobenzene, bromo-benzene and iodobenzene (55). Three factors contribute to the stability and phosphorescence of the triplet excimer.

(i)  The energy of the charge-resonance states $^3(R)_{g,u}$ is reduced due to the decrease in $(I - A)$ on halogen substitution.

(ii)  Heavy-atom substitution and the resultant increased spin-orbit coupling increase the $S_o$ - $T_q$ transition moment, $m_q$, and thus increase the splitting of the triplet exciton resonance states $^3(E_q)_{g,u}$. The heavy-atom substitution also reduces the spin-forbiddenness of the excimer phosphorescence transition.

(iii)  The $(M \parallel M)$ crystal structure maximises the halogen-halogen and $\pi$ - $\pi$ intermolecular overlap, and provides the appropriate configuration for triplet excimer formation. Values of $\Delta E_{NT} \simeq 4,800$ cm$^{-1}$ are observed in the crystal phase (53) and of $\Delta E_{NT} \simeq 8,900$ cm$^{-1}$ for the monohalobenzenes in concentrated EPA solutions at 77 K (55). There appear to be no comparative data on $\Delta E_{DM}$ for these compounds.

CONCLUSION

Excimer formation by aromatic molecules is a contest between two opposing forces, the attractive potential, $\Delta E(r)$, and the repulsive potential, $R(r)$. There are ten condensed aromatic hydrocarbons with four benzene rings or less, each with $R(r)$ of similar magnitude, but with different magnitudes of $\Delta E(r)$. In the excited singlet state they can be divided into three categories:

(i) $\Delta E(r) \gg R(r)$; anthracene and tetracene, which form photodimers;

(ii) $\Delta E(r) > R(r)$; benzene, naphthalene, pyrene, 1,2-benzanthracene, 3,4-benzophenanthrene and triphenylene, which form excimers; and

(iii) $\Delta E(r) < R(r)$; phenanthrene and chrysene, which do not form excimers.

In the excited triplet state $R(r)$ is unchanged, but $\Delta E(r)$ is greatly reduced, and it is probably that all the molecules then belong in category (iii), and do not form triplet excimers.

## REFERENCES

1. B. Stevens and E. Hutton, *Nature* <u>186</u>, 1045 (1960).

2. E.C. Lim (ed.), "Molecular Luminescence." p. 907. W.A. Benjamin, Inc., New York, 1969.

3. M.T. Vala, I.H. Hillier, S.A. Rice and J. Jortner, *J. Chem. Phys.* <u>44</u>, 23 (1966).

4. J.B. Birks, "Photophysics of Aromatic Molecules." Wiley-Interscience, New York, 1970.

5. J.B. Birks, *Chem. Phys. Letters* <u>1</u>, 304 (1967).

6. J.B. Birks and A.A. Kazzaz, *Proc. Roy. Soc. A* <u>304</u> , 291 (1968).

7. I.H. Hillier, L. Glass and S.A. Rice, *J. Chem. Phys.* <u>45</u>, 3015 (1966).

8. W. Klöpffer, in "Organic Molecular Photophysics." (J.B. Birks, ed.), Vol. 1, p. 357. Wiley-Interscience, New York, (1973).

9. W.J. Tomlinson, E.A. Chandross, R.L. Fork, C.A. Pryde and A.A. Lamola, *Appl. Opt.* <u>11</u>, 533 (1972).

10. E.A. Chandross, *J. Chem. Phys.* <u>43</u>, 4175 (1965).

11. E.A. Chandross and J. Ferguson, *ibid.* <u>45</u>, 397 (1966).

12. J.B. Birks, C.L. Braga and M.D. Lumb, *Proc. Roy. Soc. A* <u>283</u>, 83 (1965).

13. P.F. Jones and M. Nicol, *J. Chem. Phys.* <u>43</u>, 3759 (1965).

14. J.M. Thomas and J.O. Williams, *Molec. Cryst. Liq. Cryst.* <u>4</u>, 59 (1969).

15. J.O. Williams and J.M. Thomas, *ibid.* <u>16</u>, 371 (1972).

16. E.A. Chandross and C.J. Dempster, *J. Amer. Chem. Soc.* <u>92</u>, 3586 (1970).

17. J.B. Birks, *Chem. Phys. Letters* <u>4</u>, 603 (1970).

18. D. Wyrsch and H. Labhart, *ibid.* <u>8</u>, 217 (1971).

19. C.E. Swenberg and N.E. Geacintov, in "Organic Molecular Photophysics." (J.B. Birks, ed.), Vol. 1, p. 489. Wiley-Interscience, New York, 1973.

20. J.B. Birks, *Chem. Phys. Letters* 2, 417 (1968).

21. L.R. Faulkner and A.J. Bard, *J. Amer. Chem. Soc.* 91, 6495 (1969); P. Avakian, R.P. Groff, R.E. Kellogg, R.E. Merrifield and A. Suna, in "Organic Scintillators and Liquid Scintillation Counting." (D.L. Horrocks and C.-T. Peng, eds.), p. 499. Academic Press, New York, 1971.

22. J. Ferguson and A.W.-H. Mau, *Mol. Phys.* 27, 377 (1974).

23. J.B. Birks and J.B. Aladekomo, *Photochem. Photobiol.* 2, 415 (1963).

24. D.E. Applequist, E.C. Friedrich and M.T. Rogers, *J. Amer. Chem. Soc.* 81, 457 (1959).

25. R. Calas, R. Lalande and P. Mauret, *Bull. Soc. Chim. France 148* (1960).

26. F.D. Greene, *ibid.* 1356 (1960).

27. R. Calas, R. Lalande, J. Faugere and F. Moulines, *ibid.* 119 (1965).

28. O.L. Chapman and K. Lee, *J. Org. Chem.* 34, 4166 (1969).

29. D.P. Craig and P. Sarti-Fantoni, *Chem. Commun.* 742 (1966).

30. M.D. Cohen, *Molec. Cryst. Liq. Cryst.* 9, 287 (1969).

31. M.D. Cohen, Z. Ludmer, J. M. Thomas and J.O. Williams, *Chem. Commun.* 1172 (1969); *Proc. Roy. Soc. A* 324, 459 (1971).

32. A. Kawada and M.M. Labes, *Molec. Cryst. Liq. Cryst.* 11, 133 (1970).

33. E.A. Chandross and J. Ferguson, *J. Chem. Phys.* 45, 3554 (1966).

34. C.A. Parker and T.A. Joyce, *Chem. Commun.* 744 (1967).

35. T. Medinger and F. Wilkinson, *Trans. Faraday Soc.* 62, 1785 (1966).

36. W. Heinzelmann and H. Labhart, *Chem. Phys. Letters* $\underline{4}$, 20 (1969).

37. J.B. Birks, A.J.H. Alwattar and M.D. Lumb, *ibid.* $\underline{11}$, 89 (1971).

38. J.B. Birks, A.A. Kazzaz and T.A. King, *Proc. Roy. Soc. A* $\underline{291}$, 556 (1966).

39. L. Peter and G. Vaubel, *Chem. Phys. Letters* $\underline{18}$, 531 (1973); $\underline{21}$, 158 (1973).

40. O.L.J. Gijzeman, J. Langelaar and J.D.W. van Voorst, *Chem. Phys. Letters* $\underline{5}$, 269 (1970).

41. O.L.J. Gijzeman, W.H. van Leeuwen, J. Langelaar and J.D.W. van Voorst, *ibid.* $\underline{11}$, 528 (1971).

42. O.L.J. Gijzeman, Ph.D. Thesis, University of Amsterdam (1972).

43. R.B. Cundall and L.C. Pereira, *Chem. Phys. Letters* $\underline{15}$, 383 (1972).

44. R.B. Cundall and D.A. Robinson, *J. Chem. Soc. Faraday Trans. II,* $\underline{68}$, 1133 (1972).

45. D.H. Phillips and J.C. Schug, *J. Chem. Phys.* $\underline{50}$, 3297 (1969).

46. L.G. Christophorou, M.E.M. Abu-Zeid and J.G. Carter, *ibid.* $\underline{49}$, 3775 (1968).

47. J.R. Froines and P.J. Hagerman, *Chem. Phys. Letters* $\underline{4}$, 135 (1969).

48. E.A. Chandross and C.J. Dempster, *J. Amer. Chem. Soc.* $\underline{92}$, 704 (1970).

49. J. Langelaar, R.P.H. Rettschnick, A.M.F. Lambooy and G.J. Hoytink, *Chem. Phys. Letters* $\underline{1}$, 609 (1968).

50. J. Langelaar, Ph.D. Thesis, University of Amsterdam (1969).

51. R.B. Cundall and A.J.R. Voss, *Chem. Commun.* 116 (1969).

52. J. Langelaar, G. Jansen, R.P.H. Rettschnick and
    G.J. Hoytink, *Chem. Phys. Letters* $\underline{12}$, 86 (1971).

53. G. Castro and R.M. Hochstrasser, *J. Chem. Phys.* $\underline{45}$,
    4352 (1966).

54. G.A. George and G.C. Morris, *Molec. Cryst. Liq. Cryst.*
    $\underline{11}$, 61 (1970).

55. E.C. Lim and S.K. Chakrabarti, *Mol. Phys.* $\underline{13}$, 293 (1967).

# INTERSYSTEM CROSSING AND IONIC RECOMBINATION PROCESSES

## STUDIED BY PULSED LASER EXCITATION

## OF CHARGE-TRANSFER SYSTEMS

N. ORBACH AND M. OTTOLENGHI

*Department of Physical Chemistry*

*The Hebrew University*

*Jerusalem, Israel*

## INTRODUCTION

Intersystem crossing (ISC) to triplet states and ionic dissociation processes are of primary importance in determining the photophysical and photochemical consequences of excited organic charge transfer (CT) systems. New insights into the nature of these processes and into the structure of excited CT states were recently made available by the development of fast pulsed laser photolysis methods (1). By recording absorbance and conductivity changes with nanosecond time resolution, it became possible to follow directly the formation and the decay of radical ions and of excited (singlet or triplet) states.

In the present work we discuss some recently published data and present new results, aiming to reach a better understanding of ISC phenomena, as well as of ionic dissociation

and recombination processes, originating from excited intermolecular CT states. The conclusions bear on the general
problems associated with light absorption by chemical and
biological systems in which CT interactions determine the
fate of excited-states.

INTERSYSTEM CROSSING IN FLUORESCENT EXCIPLEXES (NONPOLAR

SOLVENTS) AND IN EXCITED EDA COMPLEXES (2)

EXCIPLEXES

Extensive fluorescence studies initiated by Weller and
coworkers in organic (mostly aromatic) systems (3) have shown
that in nonpolar solvents the original emission of, for
example, an electron acceptor $^1A*$, may be replaced by a new
fluorescence band attributable to the CT "exciplex", $^1(D^+A^-)*$,
formed between the excited acceptor and the quencher donor, D.
By submitting such systems to pulsed laser experiments with
nanosecond time resolution, it became possible to record the
absorption spectrum of exciplexes, providing direct evidence
for the CT nature of the fluorescent state (4). It was shown
that the exciplex absorption can be rationalized in terms of
the superimposed spectra of the radical ions $D^+$ and $A^-$, as well
as by additional transitions to localized donor and acceptor
states, $i.e.$, $^1(A^-D^+)* \xrightarrow{h\nu} (^1A*D)$ or $(A^1D*)$. Thus, in addition
to the CT nature, the spectra may provide direct information on
molecular energy levels of the donor and the acceptor,
including those which cannot be populated by direct excitation
of the isolated ground-state (A and D) or fluorescent ($^1A*$ and
$^1D*$) molecules.

Excited-state absorption spectra may also yield important
information concerning the structure of the fluorescent complex. Thus, in the case of $sym$-tetracyanobenzene (TCNB)

complexes with benzene, toluene and mesitylene, excited at
room temperature in their ground-state CT band, Mataga and
coworkers (5) have observed (in the absorption spectrum of
the fluorescent state) bands identical to those of the
dimer cation $D_2^+$. However, in rigid solutions at 77 K, the
spectrum is a simple superposition of the $A^-$ and $D^+$ absorptions.
This led them to assign the structures $^1(A^-D_2^+)*$ and $^1(A^-D^+)*$
to the emitting states in fluid media at room temperature and
at 77 K, respectively.

Early laser-photolysis experiments (6) carried out in
nonpolar solvents with anthracene and pyrene in the presence
of N,N-diethylaniline (DEA) as an electron-donor fluorescence
quencher, showed that the process is associated with substantial
population of the triplet state ($^3A*$) of the originally
excited acceptor molecule. (The corresponding triplet quantum
yields, observed for anthracene/DEA and pyrene/DEA in toluene,
are 0.55 and 0.68). An additional feature of the experimental observations was the complete lack of a growing-in
process around the $^3A*$ abosrbance maximum, which would account
for the expected evolution of the triplet state from the
thermalized exciplex, $^1(A^-D^+)*$, in competition with fluorescence emission,according to

$$^1(A^-D^+)* \longrightarrow {}^3A* + D \qquad [1]$$

The fact that in the above systems triplet evolution followed
the ( ~ 15 nsec) laser excitation profile rather than the
relatively slow decay ($\tau_{\frac{1}{2}}$ ≃ 60 nsec) of the complex fluorescence, was interpreted in terms of a fast mechanism in which
intersystem crossing occurs from nonrelaxed exciplex levels
in competition with thermalization to the emitting state,
$^1(A^-D^+)*$, which does not undergo subsequent ISC.

Ruling out reaction [1] on the basis of the lack of a
triplet absorbance growing-in, matching the fluorescence decay,

excludes the possibility that process [1] does actually occur but cannot be observed due to a coincidental mutual cancellation of the absorbances of $^3A^*$ and $^1(A^-D^+)^*$. We have thus extended the laser experiments in nonpolar solvents, investigating seven exciplex systems for which complete "initial" transient spectra, recorded 15-50 nsec from the origin of the laser pulse $(\Delta D_o)$, were compared with those recorded after the complete decay of the exciplex $(\Delta D_\infty)$ (7). The results in the region of the corresponding triplet absorption maxima, may be summarized as follows:

(i) In three cases (with methylcyclohexane as solvent), i.e., $^1TMPD^*$ + naphthalene, $^1anthracene^*$ + DMA and $^1TMPD^*$ + biphenyl (TMPD $\equiv$ N,N,N',N'-tetramethyl-$p$-phenylenediamine, DMA $\equiv$ N,N-dimethylaniline), we observed $\Delta D_\infty > \Delta D_o$, indicating an evolution of the respective triplets ($^3$naphthalene*, $^3$anthracene* and $^3$biphenyl*) from the corresponding thermalized exciplexes, in agreement with reaction [1]. The same observation for $^1anthracene^*$ + DMA was first reported by Land et al. (8). In all three cases, the initial spectrum, $\Delta D_o(\lambda)$, is substantially different from that of the triplet, $\Delta D_\infty(\lambda)$, indicating a considerable contribution of the exciplex to $\Delta D_o(\lambda)$. Although exciplex absorption spectra have been quatitatively interpreted (4), it is essentially impossible to predict exact band energies and intensities so as to permit a subtraction of the exciplex absorption from $\Delta D_o(\lambda)$ to yield the net initial triplet contribution.

(ii) In one case ($^1$pyrene* + DEA in toluene), no growing-in is observed but, again, $\Delta D_o(\lambda)$ differs from $\Delta D_\infty(\lambda)$ due to an undeterminable exciplex contribution.

(iii) In two cases (see figure 1), $^1anthracene^*$ + DEA and $^1DEA^*$ + naphthalene, $\Delta D_o(\lambda)$ is essentially identical to $D_\infty(\lambda)$ within the whole absorption range of the corresponding

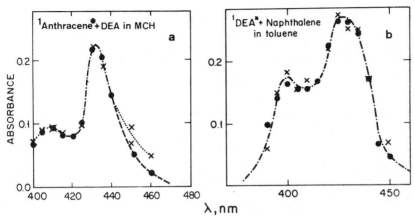

*Fig. 1 Transient absorption spectra in exciplex
systems excited by 337.1 nm, 10 nsec, $N_2$-laser pulses.
[data reproduced from ref. (7)].*
*a) $10^{-2}$ M anthracene and 0.5 M DEA in deaerated
methylcyclohexane.*
*..X.. , $\Delta D_o$ (15 nsec from the origin of the pulse)*
*..O.. , $\Delta D_\infty$ (200 nsec)*
*b) 0.2 M naphthalene and 0.25 M DEA in deaerated
toluene.*
*..X.. $\Delta D_o$ (50 nsec)*
*..O.. $\Delta D_\infty$ (200 nsec)*

triplets ($^3$anthracene* and $^3$pyrene*). Since there is no reason
to assume that the spectra of $^3$A* and $^1$(A$^-$D$^+$)* are identical,
these observations can be rationalized only in terms of a
negligible contribution of $^1$(A$^-$D$^+$)* to the absorption in the
triplet range. It therefore appears that in these cases the
initial spectrum, $\Delta D_o(\lambda)$, is solely due to triplets formed in
a fast mechanism preceding the population of the thermalized
exciplex. A similar situation is also encountered with
$^1$DEA* + biphenyl in toluene, where $\Delta D_o(\lambda) = \Delta D_\infty(\lambda)$ in the
350 - 390 nm range of the $^3$biphenyl* absorption, with the
exciplex contributing to $\Delta D_o(\lambda)$ between 390 and 425 nm.

In conclusion, although ISC from thermalized states,
competing with fluorescence emission, does occur in certain
exciplexes (case i), the experimental data in other systems

(case iii) are explainable only in terms of a fast mechanism in which triplets are populated from nonrelaxed excited complex states. At present it is impossible to evaluate the extent to which the fast process contributes to the systems of cases i and ii.

## EXCITED EDA COMPLEXES

The spectroscopy and photochemistry of EDA complexes formed between TCNB and benzene, toluene, hexamethylbenzene o mesitylene (9) bear features in common to those of the previously discussed exciplexes. Working in rigid glasses at 77 K, Mataga and coworkers (10, 11) have followed the evolution of the (CT) triplet absorption (12, 13) after laser excitation within the CT band of the above complexes. Simila: to the exciplexes of case iii, they found that the evolution of the triplet absorbance around 660 nm followed the (~ 20 nsec) pulse profile of their Q-switched ruby laser rather than that of the (~ 40 nsec) fluorescence decay. Since such observations were repeated at different wavelengths in the range of the triplet bands, they concluded that the phosphores cent state is immediately generated from the excited Frank-Condon state of the complex rather than *via* the relaxed fluorescent state.

## IONIC DISSOCIATION AND TRIPLET GENERATION FOLLOWING THE CT QUENCHING OF FLUORESCENCE IN POLAR SOLVENTS

### ROOM-TEMPERATURE DATA IN ACETONITRILE AND METHANOL

The difficulties associated with the experimental investigation of ISC following fluorescence quenching in polar solvents are, in principle, identical to those previously discussed for nonpolar solutions. Namely, instead of separating the time-dependent contributions of $^1A*$ and $^1(A^-D^+)*$, one faces the problem of separating the absorbance of $^3A*$ from those of the radical ions $^2D^+$ and $^2A^-$. These are formed in polar solvents in competition with the population (as well as from the decay) of the fluorescent exciplex (3, 14, 15). Since, as shown by chemiluminescence studies (16), the recombination of radical ions can be associated with triplet population *via* a bimolecular homogeneous process, such as

$$^2A^- + {}^2D^+ \longrightarrow {}^3A* + D \qquad [2]$$

we shall refer to $\Delta D_o(\lambda)$ and $\Delta D_\infty(\lambda)$ in polar solutions as the absorbance changes measured, respectively, before and after completion of reaction [2].

In our early laser flash photolysis studies with $^1$anthracene* + DEA and $^1$pyrene* + DEA in acetonitrile (6), we failed to observe the homogeneous recombination process [2], probably because of water contamination which efficiently competed with $^2D^+$ for $^2A^-$. It was noticed, however, that the initial absorbance, $\Delta D_o(\lambda)$, was rather high around the absorbance maximum of $^3A*$. This led to the generalization of the fast ISC mechanism to polar solutions as well. Subsequently (17), we have observed temperature effects on ISC in the above systems which were essentially independent of solvent polarity. In nonpolar solutions, the temperature effects on ISC and on the exciplex fluorescence yields and lifetimes could be rationalized only in terms of a mechanism in which triplets are generated before relaxation to the fluorescent state. The analogy between the temperature effect on ISC in polar and nonpolar solvents was considered as confirming the applicability of the prompt triplet formation path in the case of polar systems.

Along with the assignment of the initial triplet yields (observed at the end of the laser pulse) to fast ISC, we have subsequently succeeded (17) in observing the slow growing-in of $^3$pyrene* *via* the recombination of $^2$pyrene$^-$ and $^2$DEA$^+$ according to reaction [2]. We have also found that the process can be inhibited by the scavenging of $^2$pyrene$^-$ by TCNB, leading to TCNB$^-$ which regenerates the original ground-state system according to

$$TCNB^- + DEA^+ \longrightarrow TCNB + DEA \qquad [3]$$

Very recently Schomburg *et al.* (18), working in polar acetonitrile solutions of mixed pyrene/DEA/ -dicyanobenzene (DCNB) systems, have quantitatively analyzed reaction [2], establishing its second-order character. In agreement with our previous data for P/DEA/TCNB in acetonitrile, they also have observed that the generation of $^3$pyrene* *via* process [2] may be inhibited by a scavenger for $^2$pyrene$^-$ (in their case DCNB replaces TCNB in reaction [3]). They failed, however, to observe any initial triplet absorbance superimposed on the spectra of pyrene$^-$ and DEA$^+$ or (in the presence of DCNB) of DCNB$^-$ and DEA$^+$. This led them to the conclusion that fast triplet formation in electron-transfer fluorescence quenching should be ruled out as a plausible ISC path.

Being aware of the difficulties associated with establishing the triplet contribution to $\Delta D_o(\lambda)$, we have carried out a detailed analysis of laser flash photolysis data in polar (methanol or acetonitrile) solutions of pyrene or anthracene in the presence and in the absence of DCNB or TCNB. It should be pointed out that in contrast to nonpolar solutions, where no quantitative independent information for the absorption spectrum of $^1(D^+A^-)^*$ is available, the spectra of $^2D^+$ and $^2A^-$ are, in most cases, reported in the literature. In the following sections we shall use such spectra, carrying out subtraction procedures, to give the separate contribution of triplet states to $\Delta D_o(\lambda)$. It should be recalled that the accuracy of such procedures is limited by the fact that the spectra of $^2D^+$ and $^2A^-$, as available from chemical or radiation-chemical methods, are recorded under solvent and temperature conditions which differ from those of our present laser experiments. This obviously decreases the accuracy of the subtraction procedure. We shall however show: (i) that the homogeneous recombination process [2] contributes only a negligible amount to the initial spectra observed at the end of the laser pulse ($\Delta D_o(\lambda)$) and (ii) that such initial spectra contain contributions which cannot be accounted for by the superimposed spectra of $^2D^+$ and $^2A^-$ and which should be assigned to promptly formed triplet states. We now turn to the analysis of several specific systems.

## Excited Pyrene ($^1P*$) Quenched by DEA in Acetonitrile in the Presence and Absence of TCNB

Figure 2(I) shows the transient absorbance changes measured 25 nsec after firing the laser in pyrene-DEA ($2 \times 10^{-2}$ $M$) solutions in acetonitrile. A comparison with the superimposed spectra of $P^-$ and $DEA^+$ (figure 2(II), shows that the transient laser absorbance (I) exhibits, in addition to the characteristic 495 nm ($P^-$) and 465 nm ($DEA^+$) maxima, a band around 415 nm which does not belong to either $DEA^+$ or $P^-$. The separate contribution of the third band, obtained by subtracting transient (II) from (I), leads to a spectrum (figure 2, insert) which is readily identified as the characteristic 415 nm band of the pyrene triplet state. An analysis of oscillogram a in figure 3 shows that the triplet contribution observed 25 nsec after triggering cannot be due to reaction [2] which is that responsible for the relatively slow growing-in

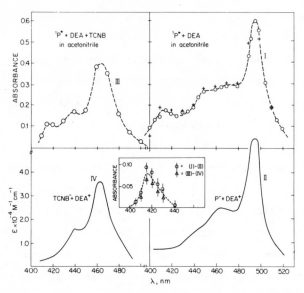

Fig. 2   Transient and difference (insert) spectra in the pulsed
$N_2$-laser photolysis of deaerated pyrene/DEA solutions in
acetonitrile.
(I)   Transient absorbance recorded 25 nsec after triggering
the laser
   (+), Sample provided by H. Staerk [P] = $10^{-4}$ M,
   [DEA] = 0.015 M, deaerated by consecutive freeze - thaw cycle
   (O), Our sample.  [P] = 2 × $10^{-5}$ M,
   [DEA] = 0.02 M deaerated by bubbling nitrogen, through sample
Appropriate corrections were carried out (see figure 3) for the
effects of scattered fluorescence light.  The absolute value
of the transient absorbance change in the second case (O) was
smaller than that for (+) by a factor of ~ 5 (due to difference
in the pyrene concentration and in the laser pulse intensity).
In the figure, the low intensity spectrum (O) was normalized
to fit the high intensity one (+).
(III)   Transient absorbance (after 400 nsec) in a sample
where 1.5 × $10^{-3}$ M TCNB was added to solution (O).
(II and IV) Superimposed (1:1) spectra of the ions.  Ionic
spectra are taken from:   a.   Ref. 18.  b. W.P. Weijland, Thesis
Free Univ. Amsterdam (1958);   W.I. Aalbersberg, G.J. Hoytink,
E.L. Mackor and W.P. Weijland, Mol. Phys. 2, 3049 (1959). c.
T. Chida and W.H. Hamill J. Chem. Phys. 44, 2369 (1966). d.
Ref.7. The same scale was used in drawing the spectra of the
ions (II,IV) and the transient absorbance changes (I,III), so
that direct substractions (insert) could be readily carried ou

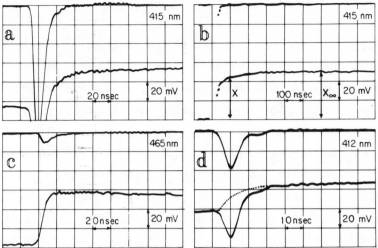

Fig. 3 *Characteristic oscillograms in the laser photolysis of pyrene/DEA in polar solvents.*
*(a) - (c), [pyrene] = $10^{-4}$ M, [DEA] = 1.5 × $10^{-2}$ M, in
acetonitrile*
*(d), [pyrene] = 1.2 × $10^{-4}$ M, [DEA] = 0.1 M, in methanol
Upper traces are in the absence of the monitoring light
beam (fluorescence only). Lower traces are in the
presence of the monitoring beam (fluorescence + absorption). The dotted lines, representing the net contribution of absorption, are obtained after correcting the
lower traces for the fluorescence contribution. (D =
log Vo/(Vo-x) and $D_\infty$ = log Vo/(Vo-$x_\infty$)).*

around 415 nm (figure 3b). Thus, a rough extrapolation of
the second-order plot, $1/(D_\infty-D)$ vs time, to the center of the
pulse (~ 8 nsec) shows that the absorbance which grows-in
during the first 25 nsec does not exceed 10 - 20% of that
attributed to the triplet (figure 2, insert). The same
conclusion is reached by an analysis of oscillogram c in
figure 3, which represents the ions at 465 nm where the absorption due to $^3P^*$ is negligible. At the initial stages of the
ions' decay ( t = 25 - 75 nsec), the drop in absorbance does
not exceed the value $\Delta D_i \simeq 0.005$ per 25 nsec. Should the
triplet absorbance observed 25 nsec after pulsing ($\Delta D_t$) be
due to reaction [2], then the relation $\Delta D_i > (\varepsilon_i/\varepsilon_t)\Delta D_t$ should
hold in which $\varepsilon_i = 2.5 \times 10^4$ $M^{-1}$ $cm^{-1}$ and $\varepsilon_t = 3.5 \times 10^4$ $M^{-1}$ $cm^{-1}$

are, correspondingly, the extinction coefficients of $^3P^*$ and
of the superimposed ions at 465 nm (19). Since the value
observed for $\Delta D_t$ (25 nsec) is ~ 0.075 (figure 2, insert), it
is evident that the above relation is far from being applicable. This rules out reaction [2] as a source of the initial
triplet absorbance at 415 nm, confirming the prompt triplet
generation. This conclusion is also consistent with the
complete insensitivity of the initial spectrum, figure 2(I),
to the magnitude of the absorbance change (see caption for the
figure), determined by the pyrene concentration and by the
laser light intensity. Should the initial (25 nsec) triplet
contribution be due to reaction [2], then it would markedly
depend on the initial concentration (*i.e.*, absorbance) of the
ions. This is in variance with figure 2(I), where the same
initial spectrum is obtained for two samples differing in
their initial absorbance change by a factor of ~3.

The presence of a triplet contribution to the initial
absorbance is also made evident by the experiments carried
out in the presence of TCNB (figure 2(III)), where the pyrene
negative ion is replaced by $TCNB^-$ formed *via* the process
$P^- + TCNB \rightarrow TCNB^- + P (\tau_{\frac{1}{2}} \simeq 30$ nsec at $[TCNB] = 1.5 \times 10^{-3}$ *M*)
(17). A subtraction of the superimposed contribution (IV) of
$TCNB^- + DEA^+$ from curve III, leads to a spectrum (figure 2,
insert) which, within the limits of our experimental accuracy,
is identical to the difference spectrum, (I) - (II), of the
previous, TCNB-free system.

### Excited Pyrene Quenched by DEA in Methanol, in the Presence and Absence of DCNB

Results and conclusions similar to the above are also
obtained when 0.015 *M* DCNB, rather than TCNB, is added to
pyrene $(10^{-4} M)/DEA$ (0.15 *M*) solutions in methanol. When the

spectra of the ions ($^1P^-$ + $DEA^+$ in the absence of DCNB, and $DCNB^-$ + $DEA^+$ in the presence of DCNB) are subtracted from the corresponding transient spectra [figure 4(Ia) and (II)], essentially the same difference spectrum, (IV), is obtained,

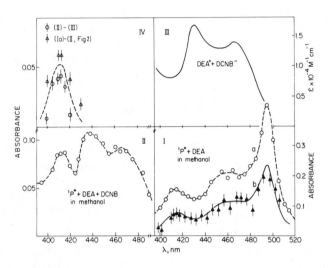

Fig. 4 *Transient and difference spectra in the pulsed $N_2$-laser photolysis of deaerated $10^{-4}$ M pyrene, 0.15 M DEA, solutions in methanol.*
*(I). Transient absorbance recorded 5 nsec (b) and 25 nsec (a) after triggering of the laser.*
*(II). Transient absorbance recorded 60 nsec after triggering of the laser in a sample where 1.5 × $10^{-2}$ M DCNB were added to the above pyrene/DEA system.*
*(III). Spectra of the superimposed ions $DEA^+$ + $DCNB^-$ (see caption for figure 2).*
*(IV). Difference spectra carried out as described in figure 2 (see figure 2(II) for the contribution of $P^-$ + $DEA^+$).*

coinciding with that of $^3P^*$. (The correction for the contribution of the ions' absorbance in the second case is very sensitive to the exact location of the 430 nm maximum of $DCNB^-$. Thus, a better agreement between the two triplet spectra in

87

figure 4(IV) is obtained if it is assumed (see figure 4) that the maximum of $DCNB^-$ in acetonitrile is slightly blue-shifted relative to that reported in the literature). This is in variance with the results of Schomburg *et al.* (18) who failed to observe any triplet contribution 50 nsec after firing the laser in a pyrene-acetonitrile solution quenched by 0.015 *M* DCNB and 0.015 *M* DEA. We should emphasize at this point that, due to the higher DEA content the residual fluorescence of $^1P^*$ in the above methanol solutions is weaker (figure 3d) than that in the previous acetonitrile solution (figure 3a). This enabled us to carry out appropriate corrections for the fluorescence scattered light, obtaining a difference spectrum [figure 4(Ib)] 5 nsec after triggering, *i.e.*, even before the completion of the laser pulse. The spectrum, which within the limits of experimental accuracy is very similar to that observed after 25 nsec, shows a clear triplet contribution which obviously cannot be attributed to reaction [2].

Moreover, should a growing-in during the first 24 nsec (according to reaction [2]) be responsible for the observed triplet contribution, it would have been considerably inhibited (down to ~ 20%) by the presence of 0.015 *M* DCNB reacting with $P^-$ with a half-life of ~ 5 nsec (18). As shown by figure 4(IV), this is obviously not the case in our experiments. Thus, triplet generation *via* reaction [2] cannot be responsible for the difference spectrum in figure 4(IV), again implying prompt triplet formation.

Excited Anthracene ($^1A^*$) Quenched by DEA in Acetonitrile (20)

Figure 5(I) shows the transient absorbance change measured 15 nsec after flashing a $10^{-3}$ *M* anthracene solution in acetonitrile quenched by 0.2 *M* DEA [data are reproduced from ref. (7)]. Curve II represents the contribution of the $DEA^+$ ion

(A$^-$ does not absorb in this spectral range). Curve III is the difference I - II and exactly coincides with the absorption of the anthracene triplet. When considering the possible contribution of reaction [2] to the triplet absorbance measured 15 nsec after triggering ($\Delta D_t$), use is again made of the expression $\Delta D_i > (\varepsilon_i/\varepsilon_t)\Delta D_t = 0.24\Delta D_t = 0.06$, where we have used $\varepsilon_i = 1.5 \times 10^4\ M^{-1}\ cm^{-1}$ and $\varepsilon_t = 6.5 \times 10^4\ M^{-1}\ cm^{-1}$. An analysis of the decay curve of DEA$^+$ at 465 nm [see figure 4 in ref. (7)], yields $\Delta D_i \sim 0.01$ per 15 nsec, for the first stages (up to 50 nsec) of the ionic decay. Thus, recalling also that it takes ~ 15 nsec to obtain the full evolution of A$^-$ and DEA$^+$ (19b), it is evident that the contribution of reaction [2] to $\Delta D_t$ (15 nsec) cannot exceed 10%.

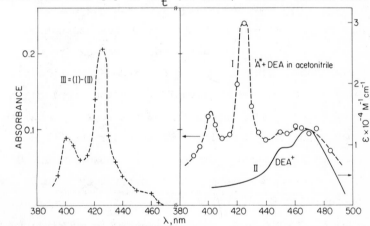

*Fig. 5  Transient and difference spectra in the pulsed N$_2$-laser photolysis of $10^{-3}$ M anthracene and 0.2 M DEA in acetonitrile.*
*(I).  Transient absorbance recorded 20 nsec after triggering (data reproduced from ref. 7).*
*(II).  Spectrum of DEA$^+$.*
*(III).  Difference (I) - (II).*

In conclusion, the present data, though limited in accuracy by some uncertainty in the exact spectra of the ions used in the subtraction procedure, unambiguously show

an initial triplet contribution which cannot be attributed to
the homogeneous recombination of ions.  Being thus in variance
with the conclusions of Schomburg *et al.* with respect to the
initial spectral changes observed after CT quenching in polar
systems (18), the data support the extension of the prompt
triplet effect to polar solutions where no exciplex fluorescence can be detected.

*OBSERVATIONS AT LOW TEMPERATURES AND IN VISCOUS SOLVENTS*

The above observations raise questions related to the
detailed mechanism of prompt triplet generation in polar
solutions.  As mentioned above, we have proposed that fast
ISC occurs from nonrelaxed exciplex states, in competition
with deactivation to the thermalized fluorescent state
(main path in nonpolar solvents) or to the solvent-shared ion
pair ($^2A^-$ ... $^2D^+$) (polar solvents).  (Mataga and coworkers
have also shown, for aromatic hydrocarbon-dimethylaniline
exciplexes and for the excited TCNB-benzene complex, that not
only ISC but also the generation of the separated $^2A^-$ and $^2D^+$
ions may occur from nonrelaxed CT states rather than from the
thermalized fluorescent complex (21)).  However, an additional
plausible source for promptly formed triplets should also
be considered.  This involves the geminate  recombination
of ions, a process which is closely associated with the nonhomogeneous initial distribution of primary irradiation
products in photochemistry and radiation chemistry known,
correspondingly, as "cage" and "spur" effects.  One particular
case of geminate recombination involves the neutralization of
a single ion pair.  The kinetics of the reaction, which in
ordinary liquids at room temperature is completed within less
than a few nanoseconds, was treated theoretically mainly by
Mozumder and coworkers (22).  Direct experimental observations

of negative ion lifetimes in radiation-chemical spurs have been reported by Thomas *et al.* in pulse irradiated cyclohexane solutions (23). No experiments have, however, been performed which lead to the direct observation of geminate neutralization of a single ion pair as it can be produced by the more selective photochemical excitation.

In view of the diffusion-controlled nature of the homogeneous process [2] in the present exciplex systems (17, 18), it is plausible that substantial geminate recombination will occur in competition with diffusional separation to the homogeneously distributed $^2A^-$ and $^2D^+$. The possibility that such a process will actually yield $^3A^*$ is associated with the overall spin configuration of the ion pair, *i.e.* $\{^2A^-(\uparrow) \ldots ^2D^+(\downarrow)\}$ recombining to singlet states, as compared to the configuration $\{^2A^-(\uparrow) \ldots ^2D^+(\uparrow)\}$ which will lead to the triplet manifold. With the purpose of obtaining a direct insight into geminate ionic processes following the excitation of CT systems, we have extended the pulsed laser experiments in polar solutions to higher solvent viscosities and to lower temperatures. At high viscosities and low temperatures, diffusion will be slowed down to an extent which may bring the geminate recombination of the ions into the time resolution of our nanosecond laser apparatus.

The characteristic pulsed laser photolysis patterns obtained for $2 \times 10^{-4}$ $M$ pyrene/0.15 $M$ DEA in methanol at low temperatures are shown in figures 6 and 7 (see also ref. 24). A gradual variation of the temperature from $20^{\circ}$ C down to $-90^{\circ}$ C slows down the bimolecular decay of the ions (table 1) as expected for a viscosity and temperature-dependent diffusion controlled process. However, in striking contrast to the room temperature behavior, the low temperature absorbance decay observed at the characteristic absorption maxima of

## TABLE 1

Temperature dependence of ionic decay rates and yields in the laser

photolysis of 2 ×10⁻⁴ M pyrene/0.15 M DEA in alcohol solutions

| Solvent | Temperature °C | Viscosity c.p. | $\tau_{\frac{1}{2}}^f$ nsec [a] | $\tau_{\frac{1}{2}}^s$ μsec [b] | $D_\infty^c$ |
|---|---|---|---|---|---|
| methanol | +16 | 0.60 | <5 | 1.5 | 0.2 |
| | −5 | 0.75 | ~9 | 1.8 | 0.16 |
| | −18 | 1.05 | ~15 | 2.5 | 0.15 |
| | −29 | 1.4 | ~18 | 3 | 0.12 |
| | −38 | 1.7 | 20 | 4 | 0.09 |
| | −77 | 5.1 | 35 | 11 | 0.04ᵈ |
| | −91 | 8.9 | 40 | 20 | 0.03ᵈ |
| % glycol in methanol-ethylene-glycol mixtures   45% | +16 | 2ᶠ | 9 | 2.5 | − |
| 80% | +16 | 6.6ᶠ | ~17 | 4.0 | − |
| isopropanol | +16ᵉ | 2.8 | 32±5 | >1 | − |
| | −77 | ~10² | 50±5 | >1 | − |

a. Half life of the fast decay measured between $D-D_\infty = 0.04$ and $D-D_\infty = 0.02$
b. Half life of the slow decay measured between $D = 0.36$ and $D = 0.018$.
c. $D_\infty$ is the residual ionic absorption measured after completion of the fast decay prior to any substantial slow decay (see figure 6).
d. Values may be excessively low due to incomplete quenching of $^1$pyrene*.
e. [pyrene] $= 4 \times 10^{-4}$ M, [DEA] $= 0.3$ M
f. at 30° C.

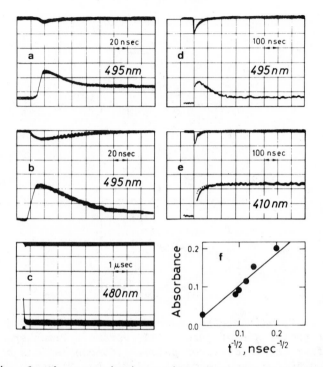

Fig. 6  Characteristic oscillograms (a-e) and kinetic
decay plot (f), in the pulsed $N_2$-laser photolysis of
pyrene/DEA in alcoholic solvents.  Upper traces are
recorded in the absence of the monitoring light
(fluorescence only);  lower traces are in the presence
of the monitoring beam (absorbance and fluorescence).
Dotted trace in e is the net absorbance contribution.
Vertical sensitivity:  20 mV/div.  Solvents and concentrations:  a.  Methanol at $-40^{\circ}C$;  b,c.  Methanol at
$-91^{\circ}C$ (in both cases [pyrene] = $2 \times 10^{-4}$ M and [DEA] =
0.15 M;  d,e.  Isopropanol at $-77^{\circ}C$ ([pyrene] = $4 \times 10^{-4}$M,
[DEA] = 0.35 M);  f.  Square root analysis of the initial
decay at 495 nm as observed in isopropanol at $+16^{\circ}C$
([pyrene] = $2.5 \times 10^{-4}$ M, [DEA] = 0.2 M).

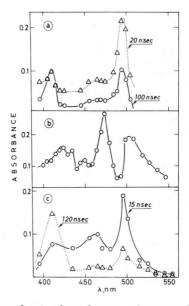

*Fig. 7   Transient absorbance changes in the pulsed $N_2$-
laser photolysis of pyrene/DEA systems.   Times specified
are from origin of laser pulse.   (Concentrations as in
figure 6).   a.   methanol at -40°C.   b.   Transient
spectrum observed 10 nsec after pulsing a pyrene/DEA
solution in toluene (reproduced from ref. 7).   Above
440 nm it represents essentially only the exciplex
absorbance.   Around 412 nm the triplet contribution may
be significant (see text).   c.   Isopropanol at -77°C.*

$P^-$ (495 nm) and $DEA^+$ (465 nm)) exhibits an additional distinct
initial stage.   For example, at -40° C the initial decay is
characterized by a half life of $\tau_{\frac{1}{2}} \simeq 20$ nsec, as compared
with $\tau_{\frac{1}{2}} \simeq 4$ μsec observed for the first half life of the slow
process.   Figure 7a shows that, in the 550 nm range, the
initial spectrum recorded 20 nsec after pulsing is essentially
identical to that recorded at 100 nsec after completion of the
first stage of the decay.   Both spectra represent the superposition of the absorbances of $P^-$ and $DEA^+$ (see the previous

analysis at room temperature), thus establishing the identification of the initially decaying species as $P^-$ and $DEA^+$. An alternative assignment of the fast decay to *e.g.*, a short-lived exciplex, is inconsistent with the extremely weak exciplex emission observed in a polar solvent such as methanol. Moreover, although reflecting the spectra of $P^-$ and $DEA^+$, the absorption spectrum of the pyrene-DEA exciplex (figure 7b) as recorded in toluene (4,7) (exhibiting intense bands at 475 nm and 505 nm with a low energy tail extending to 900 nm) is unambiguously different from that of the super-imposed ions as shown in figures 2 and 7.

Before proceeding to the analysis of the fast initial decay, attention should be paid to the low temperature absorbance patterns observed around the 415 nm range of the $^3$pyrene* absorbance maximum. As shown in figure 7a for a methanol solution at -40°C, the 100 nsec spectrum, recorded after completion of the fast decay, shows a very high contribution around 415 nm which may be assigned without ambiguity to the pyrene triplet (see above). This intense triplet band cannot arise from the homogeneous recombination [2] whose half life at -40°C is ~ $4 \times 10^{-6}$ sec, implying that only less than 5% of the ions had homogeneously reacted within the first 100 nsec. Therefore, as previously deduced from the room temperature data, a prompt triplet generation path should be invoked, in variance with the recent objections of Schomburg *et al.* (18). One should notice that the effect of fast intersystem crossing is most clearly demonstrated in these low temperature methanol solutions, where the triplet contribution is considerably higher than at room temperature (see also ref. 17). As shown in figure 7a, essentially no errors are associated in this case with the subtraction of the relatively small absorption of the ions absorption which leads to an unambiguous triplet contribution.

The same two-staged ionic decay patterns observed upon coooling methanol solutions may also be observed by increasing the solvent viscosity at room temperature. The relevant data obtained in isopropanol and in methanol-ethylene glycol mixtures are shown in figures 6 and 7 and in table 1. In all cases, the rate of the fast decay was unaffected by changes in the laser intensity when the latter was varied within a factor of ~ 10.

THE FAST INITIAL DECAY OF THE IONS

The fast ionic decay should be considered in the light of our present knowledge of nonhomogeneous initial recombination processes of radicals generated photochemically or by ionizing radiation in condensed media, known, correspondingly, as "cage" and "spur" reactions. The photochemical case is simpler in that it deals with a single radical pair where geminate recombination competes with diffusion to the bulk of the solution. (Radia-tion-chemical spurs usually contain several radical pairs). In view of their fundamental role in determining the consequence of radiation absorption in condensed media, geminate recombina-tion processes have attracted considerable attention. Theoretical treatments performed by Wijsman (25) and by Monchick (26), gave a solution for the diffusion equation of Einstein and Smoluchowski (27). The problem was also consid-ered by Noyes who approximated the diffusion process by a series of random flights (28). These treatments yielded analytical relations between the concentration of an external scavenger competing with the geminate recombination of (neutral) radicals, and the yield of the scavenging process. They did not provide, however, suitable expressions for the time-dependent concentration, $W(t)$, of the radicals (charged or uncharged) due to geminate cage recombination.

Eigen (29) has deduced expressions for the dissociation constant ($k_d$) of ion pairs. His treatment, based upon equilibrium diffusion theory, assumes the Fuoss model, according to which all ions within a contact radius (a) of the central ion are paired. Thus, in estimating the ion pair dissociation constant ($k_d = 3D\alpha/a^3$ $(\exp(\frac{\alpha}{a})-1)$ where $\alpha = e^2/$ $\epsilon kT$) in terms of the temperature, the dielectric constant ($\epsilon$) and the mean relative diffusion coefficient (D), Eigen's theory does not consider the time-dependent competition between neutralization and separation, arising from mutual diffusion at distances which are greater than a. Actually, assuming (30, 31) an equilibrium value of a $\sim$ 7 $\overset{o}{A}$ for ($^2P^-$ ... $DEA^+$) and taking D $\sim$ 6 $\times$ $10^{-6}$ $cm^2$ $sec^{-1}$ for pyrene/ DEA in methanol at $-40^{o}C$, one obtains $1/k_d \sim 10^{-9}$ nsec. This time is considerably shorter than that associated with the presently observed fast ionic decay.

Although semi-empirical expressions for $W(t)$ have been deduced for the recombination of charged or uncharged radicals in spurs (22c, 32), we shall consider the phenomena of the present work in terms of the time evolution of the neutralization of a single ion pair as treated by Mozumder (22a) which directly relates $W(t)$ with environmental parameters. Mozumder (22a) has treated the random walk of a pair of ions in their mutual coulomb field in the framework of "prescribed diffusion". The procedure solves the Smoluchowski equation (which incorporates the coulomb force between the ions and the diffusion term describing their random Brownian motion) assuming that the solution can be approximated by including, as a factor, the exact solution as derived by Einstein (27a) for neutral particles. The expression obtained for the probability

$W(r_o,t)$ that the ions will remain separate at time t, starting their random walk with a mean relative diffusion coefficient D at a distance $r_o$, is

$$W(r_o,t) = \exp(-a\{1-\text{erf}[r_o/(4Dt)^{\frac{1}{2}}]\})  \qquad [4]$$

where $\alpha = ar_o = e^2/\varepsilon kT$, is the Onsager length at which the potential energy of the ions is numerically equal to kT, and $\text{erf}(y) = 2(\pi)^{-\frac{1}{2}}\int_o^y \exp(-\xi^2)d\xi$.

This expression yields as a long-time limiting value of W

$$\lim_{(t \to \infty)} W(t) = \exp(-a) = \exp(-e^2/\varepsilon kTr_o) \qquad [5]$$

Expression [5] is identical to that derived by Onsager (33) who obtained for the first time a solution of the Smoluchowski equation in the limit $t \to \infty$.

Equation [4] implies that, taking $D = 2 \times 10^{-5}$ cm$^2$ sec$^{-1}$ and $r_o = 7$ Å for pyrene$^-$/DEA$^+$ in methanol at room temperature, only an undetectable fraction of geminate ion pairs will be left after ~ 1 nsec. However, by increasing the viscosity by a factor of at least ~ 3, or by appropriately decreasing the temperature or the dielectric constant, equation [4] predicts that it should be possible to bring a considerable fraction of the geminate recombination process to the t > 10 nsec range of our laser apparatus. It is therefore plausible to attribute the fast initial component of the ions' decay, observed in the pyrene/DEA systems of table 1, to charge annihilation within geminate ion pairs (34). Thus, the data presented in figures 6 and 7 and in table 1 indicate that the fast component becomes experimentally accessible in the t > $10^{-8}$ sec range upon

(i) Cooling methanol solutions, a process associated with an increase in viscosity. (The process is also associated with

an increase in the dielectric constant, from $\varepsilon = 34$ at $+16^{\circ}$C
to $\varepsilon = 44$ at $-40^{\circ}$C, which according to equation [4] should
partially balance the effect of lowering T and D.)

(ii)   Increasing the room temperature viscosity by increasing
the glycol content of methanol-ethylene glycol mixtures.

(iii)  Increasing the viscosity with a simultaneous decrease
in polarity when passing from methanol to isopropanol ($\varepsilon = 19$).
In agreement with the polarity effect of equation [4], the
geminate process is slower in isopropanol than in a methanol-
glycol mixture of the same viscosity.

Although such observations, along with the independency of
the decay rate on the laser intensity, are in keeping with the
identification of the fast ionic decay as being due to
geminate ion pair neutralization, they still do not provide a
test for equation [4]. A quantitative comparison between
the prediction of equation [4] and any experimental observations
require an independent estimate of the important parameter $r_o$.
Random walk may start at the equilibrium distance $\sim 7$ Å of
the thermalized ion pair (30, 31). However, thermalization
may occur at larger separations if use is made of the excess
energy initially present in the $^1$A* + D system (15). In
such a case, $r_o$ may also be affected by the solvent viscosity.
A plausible procedure leading to $r_o$ would involve the measurement of W(t = ∞), yielding $r_o$ via equation [5]. Unfortunately,
the finite width of the ($\sim 10$ nsec) laser pulse does not allow
the determination of the initial absorbance of the ions,
measured prior to any geminate decay, which is essential for
the evaluation of W(t = ∞). Attempts to slow down the decay
by increasing the viscosity are accompanied by a parallel
slowing down of the quenching of $^3$P* by DEA, with the associated
loss of time resolution. It is therefore evident that we are
observing only the secondary (slower) stages of the

99

neutralization process.

Independently of the availability of $r_o$, the limited fraction of the decay which is experimentally accessible precludes a quantitative test of equation [4] over its complete time range. It has, however, been pointed out by Mozumder that at long times W(t) of equation [4] can be approximated by the square-root law

$$W(t) \simeq W(\infty) \, \exp(\gamma t^{-\frac{1}{2}}) \simeq W(\infty) + Ht^{-\frac{1}{2}}$$

where $\gamma$ and $H$ are time-independent parameters. The same conclusion is reached by the semiempirical approach of Schuler and coworkers (32). Figure 6f shows that this relation is obeyed by the fast ionic decay of the present systems.

We may thus conclude, from the main spectroscopic and kinetic features of the fast initial decay, that we are actually directly observing the evolution of neutralization of a single ion pair. A quantitative test of the Mozumder expression, as well as of the analytical or numerical results of other treatments (23, 32), will only be possible when complete data, covering the whole range of the geminate decay, becomes available. In the bimolecular systems of the present work, the time range within which the geminate decay can be observed is limited by the finite rate of the quenching process which produces the ion pair. In order to obtain complete decay patterns, EDA complexes, such as those described by Mataga (34) where the ion pair is produced instantaneously by exciting the ground-state pair, should be investigated. Since increased viscosity may considerably decre the initial yield of the ion pair, it will be necessary to use more intense pulses of $\sim 10^{-8}$ sec duration. Alternatively the use of picosecond excitation methods may lead to the detection of geminate ionic recombinations even in low viscosity liquids at room temperature (35).

## INTERSYSTEM CROSSING MECHANISMS IN CHARGE-TRANSFER SYSTEMS

The fast ISC process, which leads to triplet states circumventing the thermalized fluorescent exciplex and the homogeneous ionic recombination, is observed for several exciplex systems in fluid (polar and nonpolar) solutions as well as for various excited EDA systems in rigid glasses. Such data are interpreted (1, 6, 7, 9, 17) by assuming that a nonequilibrium form of the CT state $^1(A^-D^+)^{\#}$, formed *via* the interaction of $^1A^*$ and D or by direct excitation of the ground-state complex $^1(AD)$, undergoes fast ISC. Triplet formation from $^1(A^-D^+)^{\#}$ competes with thermal relaxation to the fluorescent exciplex $^1(A^-D^+)^*$, as well as with ionic dissociation to the ion pair $(^2A^- \ldots {}^2D^+)$ or to the separation ions $^2A^- + {}^2D^+$. It was argued that $^1(A^-D^+)^{\#}$ is nonrelaxed with respect to both its permanent solvation structure as well as to its intrinsic vibrational energy. Fast triplet generation, following the quenching of fluorescence by molecular oxygen (36) or by inorganic salts (37, 38), has been attributed to CT interactions in short-lived excited intermolecular complexes. Such observations, along with those presently described (for which the interaction is undoubtedly CT in character), are consistent with the recognition (based on the enhancement of phosphorescence in CT complexes) that intermolecular CT interactions may provide a general mechanism for the induction of spin-orbital coupling (39). Although several mechanisms involving higher CT or locally excited triplet states have been proposed (1), the exact coupling scheme responsible for fast ISC and, especially, the dramatic effect of non-relaxed states in enhancing the process are still open questions. It is possible that the excess energy is used for crossing to higher triplet states which provide a better coupling with the $^1(A^- D^+)^{\#}$ state. Alternatively, highly distorted exciplex configurations, characterized by very effective spin-orbital coupling, may be populated before relaxation to the thermalized fluorescent state. Much further work will be required for a clarification of these problems.

The conclusion that fast ISC exclusively proceeds from prethermalized (FC) CT states (showing little sensitivity to the solvent polarity), obviously does not allow for an additional (fast) mechanism in which $^3P*$ is generated $via$ the geminate neutralization of ion pairs. Knowing, however, that the homogeneous charge annihilation process [2], does lead to pyrene triplet states, the question arises as to whether or not it is legitimate to rule out the fast, nonhomogeneous, recombination of the ions as a source for $^3P*$.

Figures 6 and 7 show that the answer to this question depends on the particular solvent involved. If the process

$$(^2P^- \ldots {}^2DEA^+) \rightarrow {}^3P* + DEA \qquad [6]$$

does actually occur, along with ground-state regeneration

$$(^2P^- \ldots {}^2DEA^+) \rightarrow P + DEA, \qquad [7]$$

one would expect to observe, exactly as for the homogeneous recombination at room temperature (17, 18), a growing-in at 415 nm kinetically matching the decay of the ions around 495 or 465 nm. Since there is no reason to believe that the relative extinction coefficients of $^3P*$ and $(^2P^- \ldots {}^2DEA^+)$ around 415 nm are significantly temperature dependent, the lack of such a growing-in in the case of low-temperature methanol solutions (figure 7a) implies that reaction [6] cannot contribute significantly to the "prompt" pyrene triplets observed prior to any substantial homogeneous neutralization (40). This situation (also prevailing in the case of methanol-glycol mixtures where the ion pair decay is detectable at room temperature) appears to be in contrast with the isopropanol system in which, as shown in figures 6 and 7c, a triplet growing-in, matching the decay of $(^2P^- \dots {}^2DEA^+)$, is actually observed.

To account for the observations in low-temperature methanol and room-temperature methanol-glycol mixtures, it is possible to argue that, if the ion pair is formed from an initially singlet system, it has a total spin quantum number of 1 so that geminate charge neutralization will favor reaction [7] over reaction [6], i.e.,

$$^1P* + DEA \rightarrow \{^2P^-(\uparrow) \dots {}^2DEA(\downarrow)\} \rightarrow P + DEA \qquad [8]$$

This implies that the spin-lattice relaxation time of the two doublet ions leading to a total triplet configuration of the pair according to

$$\{^2P^-(\uparrow) \dots {}^2DEA^+(\downarrow)\} \rightarrow \{^2P^-(\downarrow) \dots {}^2DEA^+(\downarrow)\}, \qquad [9]$$

is slower as compared with the time ($\sim 4 \times 10^{-8}$ sec) of the observed geminate recombination (in methanol at $-40^\circ C$).

In the case of isopropanol, where a fast triplet growing-in is observed, several alternative explanations may be presented:

(i)  The triplet growing-in is due to ISC within thermalized exciplexes which are formed with a low yield in isopropanol, rather than from the geminate ion pair recombination. Although the triplet growing-in occurs in the same time range in which an exciplex emission is observed (41), this possibility is inconsistent with:

a.  The fact that, for $^1pyrene* + DEA$, no ISC within the fluorescent exciplex is observed in nonpolar solvents (see above).

b.   The very low exciplex contribution to the initial absorbance decay as made evident by the failure to observe (*e.g.*, in the transient spectrum of figure 7c) bands characteristic of $^1(P^-DEA^+)*$ (figure 7b).

If the thermalized exciplex can actually be ruled out as a source for $^3P*$, then the observed triplet evolution will arise from $\{^2P^-(\downarrow) \ldots {}^2DEA^+(\downarrow)\}$. In such a case one should distinguish between the two mechanisms:

(ii)   The ion pair is initially formed in a total singlet configuration but, due to a slower geminate decay and (or) a faster spin relaxation, process [6] can efficiently compete with [7], leading to $^3P*$ + DEA *via* $\{^2P^-(\downarrow) \ldots {}^2DEA^+(\downarrow)\}$.

(iii)   Spin-lattice relaxation is much slower than geminate neutralization, but a significant fraction of ion pairs is initially produced in the overall triplet configuration. Since the ions in the present systems are formed instantaneously within less than $\sim 10^{-8}$ sec (rather than *via* the slowly decaying triplet state), this mechanism implies that ionic dissociation occurs from nonrelaxed triplets formed, *via* fast ISC, within less than $\sim 10^{-8}$ sec. This requirement is consistent with the postulated fast ISC process discussed above, as well as with the findings of Mataga and coworkers according to which ionic dissociation in excited EDA complexes may almost exclusively take place from prethermalized FC states (21).

Discrimination between alternatives (ii) and (iii) is closely associated with the spin-lattice relaxation times characterizing $^2P^-$ and $^2DEA^+$ during their mutual relative diffusion while still existing as a geminate pair. Neglecting, in a first approximation, possible magnetic interactions between the two radicals, the relaxation times should be identical to those associated with the parameter $T_1$ of the Bloch equations in paramagnetic resonance (42). The rate constant for the exponential equilibration within the doublet manifold of an, initially nonthermalized, spin system of radicals is given by $1/T_1$. $T_1$ is usually determined from the shape of ESR lines (when $T_1$ is large) or from saturation effects. Although $T_1$ values for $P^-$ and $DEA^+$ in alcoholic solvents are not available, one may refer to values in the range of $10^{-5} - 10^{-6}$ sec, which are characteristic of many organic radicals in solution. Such relaxation times appear to be too long to allow for process [9] to take place within less than $\sim 10^{-7}$ sec. However, these rough estimates of $T_1$ are certainly insufficient for definitely favoring (ii) over (iii). For such a discrimination, it will be necessary to measure $T_1$ for $P^-$ and $DEA^+$, under solvent and temperature conditions identical to those of the present laser experiments. The radical ions, produced photochemically in an ESR cavity, can be subjected to line shape and saturation effects determinations. Alternatively, by applying recently developed methods of modulated excitation (43), it may be possible to reach a time resolution which will allow the direct following of the equilibration process of a presumably nonBoltzmann initial distribution within the doublet manifold of $^2P^-$ and $^2DEA^+$ (43b, c). If ESR measurements with submicrosecond resolution will be attainable, it may also become possible to distinguish between the behavior of geminate pairs and that of homogeneously distributed ions. In any case, the presently observed CT systems may provide important and unique experimental methods for the study of spin relaxation processes in organic radical ions.

## ACKNOWLEDGEMENT

The authors wish to thank Dr. H. Levanon and Dr. D. Goodall for valuable discussions.

## REFERENCES

1.  For a recent review see:  M. Ottolenghi, *Accounts Chem. Res.* <u>6</u>, 153 (1973).

2.  In the current literature the terms "exciplex" (from excited complex - M.S. Walker, T.W. Bednar and R. Lumry, *J. Chem. Phys.* <u>45</u>, 3455 (1966)) or "heteroexcimer" are usually used for excited CT complexes formed exclusively *via* the excited state of one of the partners. The term "excited EDA (Electron-Donor-Acceptor complex") is used to describe excited complexes formed *via* the excitation of a stable ground-state CT complex.

3.  H. Leonhardt and A. Weller, *Ber. Bunsenges. Phys. Chem.* <u>67</u>, 791 (1963).

4.  R. Potashnik, C.R. Goldschmidt, M. Ottolenghi and A. Weller, *J. Chem. Phys.* <u>55</u>, 5344 (1971);  A. Alchalal, M. Tamir and M. Ottolenghi, *J. Phys. Chem.* <u>76</u>, 2229 (1972).

5.  N. Tsujino, H. Masuhara and N. Mataga, *Chem. Phys. Letters* <u>21</u>, 301 (1973);  H. Masuhara and N. Mataga, *ibid.* <u>22</u>, 305 (1973).

6.  C.R. Goldschmidt, R. Potashnik, and M. Ottolenghi, *J. Phys. Chem.* <u>75</u>, 1025 (1971).

7.  N. Orbach, J. Novros and M. Ottolenghi, *J. Phys. Chem.* <u>77</u>, 2831 (1973).

8.  E.J. Land, J.T. Richards and J.K. Thomas, *J. Phys. Chem.* <u>76</u>, 3805 (1972).

9.  N. Mataga and Y. Murata, *J. Amer. Chem. Soc.* <u>91</u>, 3144 (1969).

10. H. Masuhara, T. Tsujino and N. Mataga, *Chem. Phys. Letters* <u>12</u>, 481 (1972).

11. N. Tsujino, H. Masuhara and N. Mataga, *ibid.* 15, 357 (1972).

12. S. Iwata, J. Tanaka and S. Nagakura, *J. Chem. Phys.* 47, 2203 (1967).

13. S. Matsumoto, S. Nagakura, S. Iwata and J. Nakamura, *Chem. Phys. Letters* 13, 463 (1972).

14. H. Knibbe, D. Rehm and A. Weller, *Ber. Bunsenges. Phys. Chem.*, 72, 257 (1968).

15. N. Mataga, T. Okada and N. Yamamoto, *Chem. Phys. Letters* 1, 119 (1967); Y. Taniguchi, Y. Nishima and N. Mataga, *Bull. Chem. Soc. Japan* 45, 764 (1972); Y. Taniguchi and N. Mataga, *Chem. Phys. Letters* 13, 596 (1972).

16. A. Weller and K. Zachariasse, *J. Chem. Phys.* 46, 4984 (1967); K. Zachariasse, Ph.D. Thesis, Free University of Amsterdam, Amsterdam (1972).

17. N. Orbach, R. Potashnik and M. Ottolenghi, *J. Phys. Chem.* 76, 1133 (1972).

18. H. Schomburg, H. Staerk and A. Weller, *Chem. Phys. Letters* 21, 433 (1973); *ibid.* 22, 1 (1973).

19. The inequality sign arises from: a) The fraction ( > 25%) of ionic annihilations leading to the ground state (rather than to the triplet) of pyrene. b) The efficiency of the bimolecular recombination during the first 25 nsec, which is lower than during the next 25 nsec, due to the fact that the ions reach their final concentration only at the end of the laser pulse ( ~ 15 nsec).

20. Experiments in the presence of DCNB or TCNB, similar to those described for the pyrene systems, could not be carried out with anthracene solutions, since here the high DEA content leads to the interfering ground-state complexes (DEA·DCNB) and (DEA·TCNB). High [DEA] values

are essential for obtaining complete quenching of
anthracene solutions, since the fluorescence decay time
of $^1A*$ is shorter than that of $^1P*$ by a factor of ~ 50.

21.  Y. Taniguchi and N. Mataga, *Chem. Phys. Letters* 13, 596
     (1972);  M. Shimada, H. Masuhara and N. Mataga *ibid.*
     15, 364 (1972).

22.  a)  A. Mozumder, *J. Chem. Phys.* 48, 1659 (1968);

     b)  A. Mozumder, *ibid.* 50, 3162 (1969);

     c)  A. Mozumder, *ibid.* 55, 3026 (1971);

     d)  A.C. Abell and A. Mozumder, *ibid.* 56, 4079 (1972).

23.  J.K. Thomas, K. Johnson, T. Klippert and R. Lowers,
     *J. Chem. Phys.* 48, 1608 (1968).

24.  D.M. Goodall, N. Orbach and M. Ottolenghi, *Chem. Phys.
     Letters* 26, 365 (1974).

25.  R.A. Wijsman, *Bull. Math. Biophys.* 14, 121 (1952).

26.  L. Monchick, *J. Chem. Phys.* 24, 381 (1956).

27.  a)  A. Einstein, *Ann. Physik.* 17, 549 (1905); 19, 371
     (1906);  b)  M. von Smoluchowski, *Ann. Physik.* 21, 756
     (1906).

28.  R.M. Noyes, *J. Chem. Phys.* 22, 1349 (1954).

29.  M. Eigen, *Prog. React. Kinetics* 2, 287 (1964).

30.  H. Knibbe, Ph.D. Thesis, Free University of Amsterdam
     (1969);  K.H. Grellmann, A.R. Watkins and A. Weller,
     *J. Luminescence* 1, 2, 678 (1970).

31.  Y. Taniguchi, Y. Nishina and N. Mataga, *Bull. Chem. Soc.
     Japan* 45, 764 (1972).

32.  S.J. Rzad, P.P. Infelta, J.M. Warman and R.H. Schuler,
     *J. Chem. Phys.* 50, 5034 (1969); *ibid.* 52, 3971 (1970).

33.  L. Onsager, *Phys. Rev.* 54, 554 (1938).

34.  Very recently Irie, Masuhara, Hayashi and Mataga
     (N. Mataga, personal communication), studying the photo-
     induced ionic polymerization of α-methylstyrene by

exciting its charge-transfer complex with tetracyano-
benzene, have observed a fast decay of ions formed
from the singlet state which they attributed to
geminate recombination within ion pairs. Their observations and conclusions are in general agreement with those
of the present work.

35. An application of picosecond laser pulses leading to
the direct observation of the geminate recombination of
iodine atoms produced from the photodissociation of
$I_2$ has been recently reported. (T.J. Chuang, G.W. Hoffman
and K.B. Eisenthal, *Chem. Phys. Letters*, *25*, 201 (1974).

36. H. Tsubomura and R.S. Mulliken, *J. Amer. Chem. Soc.* *82*,
5966 (1960).

37. H. Linschitz and L. Pekkarinen, *J. Amer. Chem. Soc.* *82*,
2411 (1960).

38. A.R. Watkins, *J. Phys. Chem.* *77*, 1207 (1973).

39. S.P. McGlynn, T. Azumi and M. Kinoshita, "Molecular
Spectroscopy of the Triplet State." Prentice-Hall,
Englewood Cliffs, N.J., 1969, pp. 284-325.
N. Christodouleas and S.P. McGlynn, *J. Chem. Phys.* *40*,
166 (1964).

40. Actually, if the intial fast decay is not accompanied
by the generation of $^3P*$, one should observe a small
decay at 415 nm. (The extinction coefficient of the
pair $P^-$ + $DEA^+$ at 415 nm is ~ 20% of that at 495 nm).
Since this is not observed, the possibility that a small
fraction of the recombining ion pairs will actually lead
to $^3P*$ *via* reaction [4] cannot be definitely ruled out.
A clear answer may be provided by improving the resolution of the laser experiments and by obtaining accurate
ionic and triplet spectra in methanol at the particular

temperature involved.

41. Since the intrinsic exciplex lifetime in isopropanol is predicted to be of the order of 10 nsec, such observation (see figure 6) indicates that in addition to reactions [4] and [5] the process: $(^2P^- \ldots {}^2DEA^+) \xrightarrow{slow} {}^1(P^-DEA^+)* \xrightarrow{fast} P + DEA + h\nu$ does also take place. (See papers by Bard, Weller and Zachariasse in the present volume).

42. See for example P.B. Ayscough in "Electron Spin Resonance in Chemistry.", Methuen & Co. Ltd., London, 1967.

43. a)  H. Levanon and S.I. Weissman, *J. Amer. Chem. Soc.* **93**, 4309 (1971).

    b)  P.W. Atkins, R.C. Gurd, K.A. McLauchlan and A.F. Simpson, *Chem. Phys. Letters* **8**, 55 (1971).

    c)  J.B. Pedersen and J.H. Freed, *J. Chem. Phys.* **58**, 2746 (1973); **59**, 2869 (1974).

ELECTRONIC STRUCTURES AND DYNAMICAL BEHAVIOR

OF SOME EXCIPLEX SYSTEMS

N. MATAGA

Department of Chemistry, Faculty of Engineering Science

Osaka University, Toyonaka, Osaka, Japan

INTRODUCTION

Recently, detailed studies on the structures and dynamic behavior of excited CT systems have been carried out by means of nsec time-resolved fluorescence measurements as well as by transient absorption and photoconductivity measurements with the nsec laser flash photolysis method. These sorts of investigations have provided much insight into the nature of the CT interactions in excited EDA complexes (1 - 5) such as excited 1,2,4,5-tetracyanobenzene(TCNB)-aromatic hydrocarbon complexes; similar studies have been made on exciplex systems (6). The considerable polar nature of some typical heteroexcimers can be demonstrated by measuring the effect of solvent polarity on the frequency, the decay time and quantum yield of fluorescence as well as by the $S_n \longleftarrow S_1$ absorption spectral measurements using the laser flash photolysis method.

Upon increasing the solvent polarity, both the fluorescence quantum yield and decay time of some typical

113

heteroexcimers such as pyrene-dimethylaniline(DMA) and anthra-
cene-diethylaniline(DEA) systems decrease. The fluorescence
yield decreases more rapidly with increasing solvent polarity
than does the corresponding fluorescence lifetime. This fact
was recognized in the early investigations on the hetero-
excimer (7, 8) and a detailed model to explain the observations
was required. There are two interpretations for these observations.

(i) Competition of two processes $k_{IP}$ and $k_c$ in the
encounter complex A*. . . D (7).

$$A^* + D \longrightarrow A^* \cdots D \underset{k_{IP}}{\overset{k_c}{\rightleftarrows}} \begin{array}{l} (A^-D^+)^* \\ \downarrow k_I \\ A_s^+ \cdots D_s^+ \xrightarrow{k_d} A_s^- + D_s^+ \end{array}$$

$k_{IP}$ and $k_I$ increase in polar solvents. $(A^-D^+)^*$ is the
fluorescent exciplex and $A_s^- \cdots D_s^+$ is the nonfluorescent
solvent-shared ion pair which subsequently dissociates into
ions.

(ii) Solvent-induced change of the electronic and
geometrical structure of the exciplex (8).

The exciplex state may be considered as a resonance
hybrid of the electron transfer configuration $(A^-D^+)$ mixed
with the locally excited configuration $(A^*D)$ or $(AD^*)$. On
the one hand, the mixing of these configurations leads to
the increase of the binding energy, but on the other hand,
this interaction decreases the electric dipole moment of the
complex. The solvation of the exciplex by the polar solvents,
however, will oppose the electronic delocalization tendency
since the solvation energy increases with increasing dipole
moment of the exciplex. Thus, the exciplex electronic
structure may become more polar with increasing polarity of
the solvent, and this change of the structure leads to the

decrease of the radiative transition probability. Contrary
to this, the radiationless transition probability increases
with solvent polarity because the energy gap between the
relevant electronic levels becomes smaller in a more polar
solvent and, furthermore, because of the ionic dissociation
of the complex in a more-or-less polar solvent.

According to the investigations made up to now, the
solvent-induced electronic structure change does not appear
to be important for the interpretation of the effect of
solvent polarity upon the fluorescence yield and decay time
of such strongly polar exciplexes as anthracene-DEA and
pyrene-DMA systems. However, such an effect seems to exist
in principle and may be of some importance in general for
the interpretation of electronic structures of molecular
complexes in solution since they have more labile structures
than the ordinary molecules. We would like to discuss the
actual examples of the solvent-induced electronic structure
change as well as some other recent results of our studies
on the dynamic processes of exciplexes.

In order to examine the geometrical requirements for the
formation of the fluorescent exciplex state and for the
fluorescence quenching due to the electron transfer, we have
studied the excited states of various intramolecular exciplex
systems of the following types:

where the aromatic hydrocarbon is anthracene or pyrene (9).
Several of these intramolecular exciplex systems show clearly
the solvent-induced change of electronic structure during its
lifetime.

SOLVENT INDUCED CHANGES OF EXCIPLEX ELECTRONIC STRUCTURES (10)

In this section , we will discuss the systems anthracene-
$(CH_2)_n$⟨○⟩—$N(CH_3)_2$ (n=0,1,2) and pyrene-$(CH_2)_n$⟨○⟩—$N(CH_3)_2$
(n=1) (abbreviated as $(I)_{n=0,1,2}$ and $(III)_{n=1}$, respectively).
Absorption spectra of $(I)_{n=1,2}$ and $(III)_{n=1}$ do not show
any indication of ground-state interaction. In contrast to
this, the absorption spectra of $(I)_{n=0}$ are a little broadened
compared to those of the components, indicating a weak ground-
state interaction. The longest wavelength absorption band of
$(I)_{n=0}$ may be a superposition of the anthracene band a little
modified by the substitution and a very weak CT band. (In
the case of acetonitrile solution, we have observed a very
weak band at about 500 nm in the excitation spectra. This
band might be ascribed to the CT absorption that has become
noticeable in a strongly polar solvent.)

The fluorescence wavenumbers $(\tilde{\nu}_f)$ of these compounds were
plotted against $[2(\varepsilon-1)/(2\varepsilon+1) - (n^2-1)/(2n^2+1)]$ where $\varepsilon$ and
n are the dielectric constant and refractive index of the
solvent, respectively. Some examples are shown in figure 1.
The plot for $(I)_{n=0}$ is not a straight line; this result indicates a solvent-induced change of the excited electronic
structure. The change in the electronic structure can also
be concluded from the solvent dependence of the fluorescence
yield and lifetime. From the decrease of the fluorescence
yield and the increase in its lifetime, we conclude there is

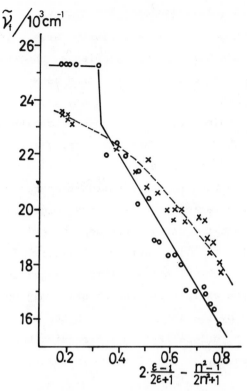

Fig. 1   The observed CT fluorescence wavenumber ($\tilde{\nu}_f$) vs $[2(\varepsilon-1)/(2\varepsilon+1) - (n^2-1)/(2n^2+1)]$ relations for $(I)_{n=0}$ and $(I)_{n=1}$.

a decrease in the fluorescence radiative transition probability, $k_f$, in the polar solvents (9).   Other evidence for the solvent-induced electronic structure change comes from the $S_n \longleftarrow S_1$ absorption spectra.   The $S_n \longleftarrow S_1$ spectra in cyclohexane solution are somewhat similar to those of 9-phenylanthracene, but the spectra in ether and acetonitrile are rather similar to those of anthracene anion (9).

The fluorescence spectra of $(I)_{n=1,2}$ and $(III)_{n=1}$ in non-polar solvents can be ascribed to the anthracene or pyrene part;   they show the CT fluorescence in polar solvents.   The CT fluorescence appears at the value of the polarity parameter

$[2(\varepsilon-1)/(2\varepsilon+1) - (n^2-1)/(2n^2+1)] \simeq 0.3$ where the red shift
with the solvent polarity is especially large. In comparison
with the case of sandwich-type exciplexes, the stabilization
of the CT state due to the coulomb force between the pair may
be smaller in these cases because of the larger separation
between the two moieties. Thus, the electron-transfer state
is placed at higher energy than the locally-excited (LE) state
$(S_1)$ of anthracene or pyrene. There may arise only a very
small interaction between the electron-transfer configuration
and the LE configuration because of the large distance between
the two groups and the unfavorable geometrical configuration
for the delocalization interaction. However, since the
dipole moments in the CT state are much larger than those of
the sandwich-type exciplexes, the CT state may be stabilized
rather strongly by the interaction with the polar solvent.
If we neglect the delocalization interaction between the two
moieties, the energy of the CT state in solution may be given
by

$$E_{CT} = I - A - C - \frac{\mu_e^{eq2}}{a^3}\left(\frac{\varepsilon-1}{2\varepsilon+1}\right)$$

$$= I - A - C - \frac{1}{2}\mu_e^{eq2}f_\varepsilon \qquad [1]$$

where I and A are the ionization potential of the donor and
electron affinity of the acceptor, respectively. C is the
coulomb interaction energy between the donor cation and
acceptor anion calaculated by using the MO's putting point
charges at each AO. a is the cavity radius in the Onsager's
reaction field theory and $\vec{\mu}_e^{eq}$ is the dipole moment of the
excited equilibrium state. The value of $(\mu_e^{eq3}/a^3)$ for $(I)_{n=1}$
was estimated to be ~ 2.1 e$V$ from the $\tilde{\nu}_f \sim [2(\varepsilon-1)/(2\varepsilon+1) -$
$(n^2-1)/(2n^2+1)]$ relation. Taking I = 7.5 e$V$, A = 0.5 e$V$,

$C = 3.2$ eV and $(\varepsilon-1)/(2\varepsilon+1) \simeq 0.26$ (at values of $(\varepsilon-1)/(2\varepsilon+1)$ larger than 0.26, CT fluorescence appears), $E_{CT} \simeq 3.25$ eV. Since the ground state will have no dipole moment, the Franck-Condon ground state (with respect to the dipole solvation) may be destabilized by

$$\frac{\mu_e eq^2}{a^3} \left( \frac{\varepsilon-1}{2\varepsilon+1} - \frac{n^2-1}{2n^2+1} \right) = \frac{1}{2}\mu_e eq^2 (f_\varepsilon - f_n) \qquad [2]$$

Since this amounts to $\sim 0.12$ eV, $\tilde{\nu}_f = (E_{CT} - 0.12)$eV$= 3.13$ eV. This $\tilde{\nu}_f$ value is almost equal to that of anthracene LE transition. Thus, in the case of $(I)_{n=1}$, the solvent-induced change of the electronic structure can be regarded as a solvation-induced reversal of two energy levels which do not interact with each other. The situation in the case of $(I)_{n=2}$ and $(III)_{n=1}$ is quite analogous.

This sort of behavior of the fluorescence spectra is not limited to the above compounds. The observation (11) that the fluorescence spectrum of 9,9'-dianthracene depends considerably on the solvent polarity, showing a red shift as the polarity is increased, seems to be a special case of such a solvent-induced change of electronic states (12, 13). Recently, we have examined the solvent effect on the fluorescence lifetime and $S_n \leftarrow S_1$ spectra of this compound by using a mode-locked ruby laser (14). The width of the laser pulse was $ca.$ 20 psec, measured by the two photon fluorescence method. We have excited the solution by means of a single pulse taken out from the mode-locked pulse train by a single pulse selector. The observed fluorescence lifetime in $n$-hexane is 8 nsec while it is more than 20 nsec in polar solvents such as ethanol and acetone; moreover, the fluorescence yield decreases with an increase in the solvent polarity.

These two facts indicate that there is a change of the
electronic structure induced by the interaction with polar
solvents. We have observed the $S_n \longleftarrow S_1$ spectrum of 9,9'-
dianthracene by using the self-phase-modulation or psec
continuum light as the spectroflash. The psec continuum light
pulse was generated by focusing the single pulse into several
substances including water. A schematic diagram of the
optical arrangement of the apparatus is indicated in
figure 2. The $S_n \longleftarrow S_1$ spectra of dianthracene observed in

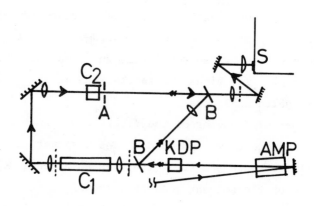

*Fig. 2 Schematic diagram of the optical arrangement
for the measurement of $S_n \longleftarrow S_1$ spectra by psec laser
flash photolysis.*
*AMP: amplifier for the single pulse taken out from
the mode-locked pulse train.*
*B: mirror reflecting the 347.2 nm light and trans-
mitting the visible light.*
*———▶: 694.3 nm light.      ——▶: 347.2 nm light.*
*➡: psec continuum light. ▌ : 100% reflecting mirror.*
*$C_1$: cuvette containing the substance to generate the
psec continuum light.*
*$C_2$: curvette containing the solution for measurement.*
*A: aperture.         S: spectrograph.*

n-hexane and acetone are shown in figure 3, where the
corresponding $S_n \longleftarrow S_1$ spectra of anthracene are given for

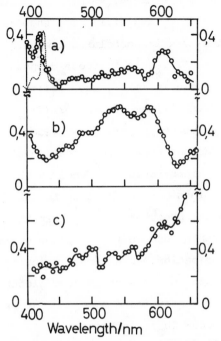

Fig. 3   The $S_n \longleftarrow S_1$ spectra of 9,9'-dianthracene
and anthracene observed by means of psec laser flash
photolysis.   The ordinate is approximately proportional to the absorbance.
a)   anthracene in n-hexane.   The dotted curve represents the triplet-triplet absorption band.
b)   9,9'-dianthracene in n-hexane.
c)   9,9'-dianthracene in acetone.

the purpose of comparison.   The spectra were taken at the

delay time of 500 psec from the exciting pulse.   The spectrum

of dianthracene in n-hexane is not the same as that of

anthracene but they are nevertheless somewhat similar.

However, the spectrum observed in acetone is quite different

from that in n-hexane, and somewhat similar to that of

anthracene ion.   This is also a direct demonstration of the

solvent-induced electronic structure change, just as we have

demonstrated in the case of $(I)_{n=0}$   (9).

In addition to the above, we have also studied $(I)_{n=3}$ and $(III)_{n=2,3}$. $(I)_{n=3}$ and $(III)_{n=2,3}$ show an exciplex fluorescence band even in nonpolar solvents. This result indicates that the CT levels in these compounds are considerably lower than those in $(I)_{n=1}$ and $(III)_{n=1}$. The $\tilde{\nu}_f$ $[2(\varepsilon-1)/(2\varepsilon+1) - (n^2-1)/(2n^2+1)]$ relations for $(I)_{n=3}$ and $(III)_{n=2,3}$ throughout all values of polarity parameter $[2(\varepsilon-1)/(2\varepsilon+1) - (n^2-1)/(2n^2+1)]$ do not show any indication of appreciable change of electronic structure induced by the interaction with polar solvents. This may be the case because the CT states of these compounds are presumably lower than the LE states even in nonpolar solvents, due to the stronger coulomb interaction between the donor cation and acceptor anion in the CT state.

In relation to the solvent-induced electronic structure change, we have recently found two peculiar examples, 1,1'- and 9,9'-anthracene-$(CH_2)_2$-anthracene (15). The fluorescence spectra of these compounds show a dependence on the solvent polarity. In nonpolar solvents, the spectra are quite similar to those of anthracene. However, in polar solvents, a broad band appears at slightly longer wavelength than the anthracene band. The broad fluorescence band of the 1,1'-compound has a lifetime of approximately 100 nsec and may probably be ascribed to an exciplex state. Because the extent of the red shift of the exciplex band with increase of the solvent polarity is very small, the exciplex fluorescent state does not seem to have a polar electronic structure. Nevertheless, the solvent polarity is quite effective in the formation of the exciplex state. The exciplex state might be formed easily *via* a CT state, but its direct formation from the LE state may not be as easy, requiring some activation energy. Actually, we have observed a temperature effect on the formation of the exciplex state. Detailed studies for the elucidation of this phenomenon are now going on in our laboratory.

In addition to above studies, we have made a detailed investigation of the solvent shift of intermolecular exciplex fluorescence of the 1-cyanonaphthalene-naphthalene system, which clearly shows an increase of the exciplex dipole moment with the increase in solvent polarity, as indicated in figure 4.

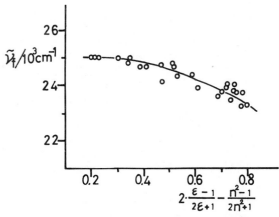

Fig. 4  The observed $\tilde{\nu}_f$ vs $[2(\varepsilon-1)/(2\varepsilon+1) - (n^2-1)/(2n^2+1)]$ relation for the 1-cyanonaphthalene-naphthalene exciplex.

The above experimental results may be summarized on the basis of the theoretical model which is an extension of a previous one (16). Namely, we assume equation [3] for the wavefunction of the exciplex system.

$$\Psi = C_1 \Psi(A^- D^+) + C_2 \Psi(A^*D) + C_3 \Psi(AD) \qquad [3]$$

The first term represents the lowest CT configuration, the second term is the locally excited ('La) configuration and the last term is the ground-state configuration. We approximate here the solute-solvent interaction by a simple continuum model using Onsager's reaction field. Then, the hamiltonian of the system may be given by

$$\mathcal{H} = \mathcal{H}_o + \mathcal{H}' = \mathcal{H}_o - \vec{\mu}_{op} \cdot \vec{F} = \mathcal{H}_o - \vec{\mu}_{op} \cdot \vec{\mu} \, f_\varepsilon \qquad [4]$$

where $\mathcal{H}_o$ is the hamiltonian of the exciplex in the absence of the solvent, $\vec{\mu}_{op}$ is the dipole operator, $\vec{F}$ is the Onsager's reaction field, $f_\varepsilon$ is the polarity parameter and $\vec{\mu}$ may be given by

$$\vec{\mu} = C_1^2 <\Psi (A^- D^+) | \vec{\mu}_{op} | \Psi (A^- D^+)> \ = C_1^2 \, \vec{\mu}_o \qquad [5]$$

We denote the matrix elements of the hamiltonian as follows

$$<\Psi (A^- D^+) | \mathcal{H} | \Psi (A^- D^+)> = <\Psi (A^- D^+) | \mathcal{H}_o | \Psi (A^- D^+)> + <\Psi (A^- D^+) | \mathcal{H}' | \Psi (A^- D^+)>$$

$$= E_c - <\Psi (A^- D^+) | \vec{\mu}_{op} \cdot \vec{F} | \Psi (A^- D^+)>$$

$$= E_c - C_1^2 \cdot \mu_o^2 \, f_\varepsilon \qquad [6]$$

$$<\Psi (A^- D^+) | \mathcal{H} | \Psi (A*D)> \approx <\Psi (A^- D^+) | \mathcal{H}_o | \Psi (A*D)> = \alpha \qquad [7]$$

$$<\Psi (A^- D^+) | \mathcal{H} | \Psi (A \ D)> \approx <\Psi (A^- D^+) | \mathcal{H}_o | \Psi (A \ D)> = \beta \qquad [8]$$

$$<\Psi (A*D) | \mathcal{H} | \Psi (A*D)> \approx <\Psi (A*D) | \mathcal{H}_o | \Psi (A*D)> = E_e \qquad [9]$$

$$<\Psi (A \ D) | \mathcal{H} | \Psi (A \ D)> \approx <\Psi (A \ D) | \mathcal{H}_o | \Psi (A \ D)> = E_g \qquad [10]$$

By putting equations [3] and [6] - [10] into the Schrödinger equation, $\mathcal{H}\Psi = E\Psi$, we are led to

$$C_1 (E_c - C_1^2 \, \mu_o^2 \, f - E) + C_2 \alpha + C_3 \beta = 0$$

$$C_1 \alpha + C_2 (E_e - E) = 0$$

$$C_1 \beta + C_3 (E_g - E) = 0 \qquad [11]$$

The solution can be obtained by solving equation [12] for various values of polarity parameter $f_\varepsilon$.

$$\{E_c - E - \frac{\alpha^2}{(E_e - E)} + \frac{\beta^2}{E}\} \{1 + \frac{\alpha^2}{(E_e - E)^2} + \frac{\beta^2}{E^2}\} = \mu_o^2 \cdot f_\varepsilon \qquad [12]$$

We denote the lowest and the 2nd lowest eigenvalues obtained from equation [12] as $E_o$ and $E_1$, respectively. The energies of the excited equilibrium state $(E_e^{eq})$ and the Franck-Condon ground state $(E_g^{FC})$ may be given by

$$E_e^{eq} = E_1 + \frac{1}{2} \mu_e^{eq} f_\varepsilon$$

$$E_g^{FC} = E_o - \vec{\mu}_g^{FC} \cdot \vec{\mu}_e^{eq} + \frac{1}{2} \mu_e^{eq^2} (f_\varepsilon - f_n) - \frac{1}{2} (\mu_g^{FC})^2 f_n \qquad [13]$$

where $\vec{\mu}_g^{FC}$ is the dipole moment of the Franck-Condon ground state. When $\vec{\mu}_g^{FC}$ can be neglected, the exciplex fluorescence wavenumber is given by

$$hc\tilde{\nu}_e = E_1 - E_o + \frac{1}{2}(\mu_e^{eq})^2 f_n \qquad [14]$$

The solvent shift of the fluorescence of $(I)_{n=1}$, for example, can be well reproduced by means of the following parameters: $E_c = 3.70$ eV, $E_e = 3.14$ eV, $\vec{\mu}_o = 28.0$ D, $\alpha = 0.05$ eV and $\beta = 0$. Since $\beta$ can be neglected and $\alpha$ is very small, the situation in this case can be regarded as a solvation-induced reversal of two energy levels. The same conclusion can be obtained by the more simplified procedure as mentioned before. A similar result was also obtained for $III)_{n=1}$; however, in the case of $(I)_{n=0}$, we cannot neglect but must use fairly large values both for $\alpha$ and $\beta$. The observed solvent shift of $(I)_{n=0}$ can be reproduced satisfactorily by the following parameter values: $E_e = 3.1$ eV, $E_c = 3.6$ eV, $\vec{\mu}_o = 20.2$ D, $\alpha = 0.4$ eV and $\beta = 0.3$ eV. The calculated $\vec{\mu}_e^{eq}$ shows a gradual increase with an increase in the solvent polarity.

In the case of the 1-cyanonaphthalene-naphthalene system, however, we must take into consideration the four configurations, $A^-D^+$, $A^+D^-$, A*D and AD*, since they are rather close to each other.

$$\Psi = C_1\Psi(A^-D^+) + C_2\Psi(A^+D^-) + C_3\Psi(A*D) + C_4\Psi(AD*) \qquad [15]$$

The dipole moment of the exciplex fluorescent state may be given by

$$\vec{\mu} = (C_1^2 - C_2^2)\vec{\mu}_o \qquad [16]$$

We denote the necessary matrix elements as follows

$$\langle\Psi(A^-D^+)|\mathcal{K}_o|\Psi(A^-D^+)\rangle = E_1$$

$$\langle\Psi(A^+D^-)|\mathcal{K}_o|\Psi(A^+D^-)\rangle = E_2$$

$$\langle\Psi(A*D)|\mathcal{K}_o|\Psi(A*D)\rangle = E_A = 3.94 \text{ eV}$$

$$\langle\Psi(A\ D*)|\mathcal{K}_o|\Psi(A\ D*)\rangle = E_D = 4.33 \text{ eV}$$

$$\langle\Psi(A^-D^+)|\mathcal{K}_o|\Psi(A*D)\rangle \approx \langle\Psi(A^+D^-)|\mathcal{K}_o|\Psi(A\ D*)\rangle = \alpha$$

$$\langle\Psi(A^-D^+)|\mathcal{K}_o|\Psi(A\ D*)\rangle \approx \langle\Psi(A^+D^-)|\mathcal{K}_o|\Psi(A*D)\rangle = \gamma$$

$$\langle\Psi(A^-D^+)|\mathcal{K}'|\Psi(A^-D^+)\rangle = -(C_1^2 - C_2^2)\mu_o^2\ f_\varepsilon = -X$$

$$\langle\Psi(A^+D^-)|\mathcal{K}'|\Psi(A^+D^-)\rangle = (C_1^2 - C_2^2)\mu_o^2\ f_\varepsilon = X \qquad [17]$$

where $E_A$ and $E_D$ are observed energies of $S_1$ state of 1-cyanonaphthalene and naphthalene, respectively. In terms of these matrix elements, the nonlinear secular equation is written as

$$C_1(E_1 - X - E) + C_3\alpha + C_4\gamma = 0$$

$$C_2(E_2 + X - E) + C_3\gamma + C_4\alpha = 0$$

$$C_1\alpha + C_2\gamma + C_3(E_A-E) = 0$$

$$C_1\gamma + C_2\alpha + C_4(E_D-E) = 0 \qquad\qquad [18]$$

We solve the secular equation for the gas phase (X=0) and obtain the dipole moment $\vec{\mu} = (C_1^2-C_2^2)\vec{\mu}_o$, from which we can evaluate X. With this X value, we start the iterative calculation, assuming $\vec{\mu}_o = 16$ D, a = 5.5 Å. The observed solvent shift can be approximately reproduced by using the following values for the energies: $E_1 = 3.8$ eV, $E_2 = 4.1$ eV and $\alpha$, $\gamma \approx 0.4 - 0.5$ eV. The calculated result indicates that the CT character of the exciplex state changes from 30% in non-polar solvent to ca. 70% in strongly polar solvent such as acetonitrile. The solvent-induced electronic structure change may finally lead to the ionic dissociation of the intermolecular exciplex in a strongly polar solvent. However, for ionic dissociation to occur, there must arise a more extensive solvation leading to the separation of the donor cation and acceptor anion, rather than the mere solvation of the fluorescent exciplex state.

CHARGE-TRANSFER AND HYDROGEN ATOM TRANSFER REACTIONS IN

EXCITED AROMATIC HYDROCARBON-ANILINE SYSTEMS

As we have reported previously (17, 18), when aniline, N-methylaniline and naphthylamines are used as electron donors for exciplex formation with aromatic hydrocarbons in solution at room temperature, contrary to the case where the donors are DMA or DEA and N,N-dimethylnaphthylamines, we cannot observe exciplex fluorescence but only a quenching of the aromatic hydrocarbon fluorescence, both in nonpolar and polar solutions. It seems plausible that charge-transfer

interaction in the encounter complex followed by some
efficient radiationless processes plays an important role in
the case of the quenching by amines with a free N-H group.
Therefore, if we can stabilize the complex in a rigid matrix
at low temperature, it may be possible to observe the exciplex
fluorescence, even in these cases. It was confirmed by us
previously that we can easily produce the loose ground-state
complex of an aromatic hydrocarbon with DMA in a rigid cyclohexane matrix at low temperatures (17). When excited by
light absorption, this loose complex easily shifts to the
fluorescent exciplex state. We have actually examined
pyrene-aniline and pyrene-N-methylaniline systems in a cyclohexane matrix at low temperatures and observed the exciplex
fluorescence band in approximately the same wavelength range
as that of the pyrene-DMA exciplex. However, the fluorescence
intensity of the pyrene-aniline or pyrene-N-methylaniline system is much smaller than that of the pyrene-DMA system. This
result seems to indicate that, even at low temperatures in the
rigid matrix, radiationless processes in the former are predominant.

In order to elucidate the mechanism of the quenching of
the aromatic hydrocarbon fluorescence by amines with a free
N-H group, we are now conducting detailed studies on several
systems by means of fluorescence measurements as well as by
transient absorption measurements with laser flash photolysis
(19).

Since we have observed weak exciplex fluorescence for
the pyrene-aniline system at low temperatures in the rigid
matrix, the radiationless processes seem to be competing
with the exciplex fluorescence transition. According to our
recent measurements, the quenching rate constant $k_q$ for the

pyrene-aniline system at room temperature ($20^{\circ}$ C) was as
follows: $0.2 \times 10^9$ $M^{-1}$ $sec^{-1}$ in methylcyclohexane,
$0.5 \times 10^9$ $M^{-1}$ $sec^{-1}$ in n-hexane, $2.6 \times 10^9$ $M^{-1}$ $sec^{-1}$ in
isopropyl alcohol and $5.3 \times 10^9$ $M^{-1}$ $sec^{-1}$ in acetonitrile.
The above values of the quenching rate constant are somewhat
smaller compared to those of the exciplex formation or the
fluorescence quenching reaction of pyrene-DMA or pyrene-DEA
system (20). The $k_q$ values for the pyrene-aniline system
are rather close to those of the pyrene-aliphatic amine system.
These facts seem to indicate that an additional activation
energy over that for the diffusional motion appears necessary
for the quenching process in the pyrene-aniline system both
in nonpolar and polar solvents. Presumably, additional activation energy over that for the diffusional motion may be needed
for the charge transfer in the encounter complex. Although
the additional activation energy may be ascribed to the change
of the structure of amine in the case of the aliphatic amine
donors, such an effect may be small in the case of aniline.
However, one should note that the ionization potential of
aniline is about 0.6 eV higher than that of DMA, which might
lead to an additional activation energy for the charge transfer
interaction in the encounter complex.

One of the possible mechanisms for this quenching process in the complex may be the charge transfer followed by a
rapid proton transfer.

$$A^* \ldots H\text{-}D \longrightarrow (A^- \ldots H\text{-}D^+)^* \longrightarrow \dot{A} - H \ldots \dot{D} \longrightarrow [19]$$

Or, it may be a hydrogen atom transfer process assisted by
a partial charge transfer in the complex.

$$A^* \ldots H\text{-}D \longrightarrow (A^{\delta-} \ldots H\text{-}D^{\delta+})^* \longrightarrow \dot{A} - H \ldots \dot{D} \longrightarrow$$
$$[20]$$

The radical pair $\dot{A}$-H $\ldots$ $\dot{D}$ will partly dissociate into

(A-H+D) and partly go back to the ground state (A+HD). A reaction mechanism similar to [19] was assumed by Yang (20) for the primary process in the photochemical addition of secondary amines to anthracne.

There have been numerous investigations of the photo-reduction of carbonyl compounds (22). Particularly active studies have been made on the photochemical behavior of ketones in the triplet state in order to elucidate the mechanism of their photoreduction (23, 24). In the cases of hydrogen abstraction by triplet ketones, especially for the ketone-amine systems, it is frequently assumed that the photoreduction proceeds *via* charge transfer from the donor to the triplet ketone followed by proton transfer. In order to elucidate further the reaction mechanism, more-or-less direct nsec laser flash photolysis studies have been carried out on the benzophenone-amine systmes in various solvents of different polarities (25, 26). In the case of tertiary aromatic amines, it has been confirmed that there is almost no ketyl radical formation; rather, complete ionic photo-dissociation to benzophenone anion and amine cation occurs in a strongly polar solvent such as acetonitrile, while photoreduction due to the hydrogen atom transfer from amine to ketone occurs efficiently in nonpolar solvents. Moreover, it has been found that these two processes compete with each other, depending upon the solvent polarity (25, 26). The reaction mechanism in these cases seems to be as follows. The charge-transfer interaction from amine to the triplet ketone in the encounter complex is followed by the solvation to the separated ion radicals as well as by the hydrogen transfer process assisted by the charge-transfer interaction. In a strongly polar solvent, the solvation process to the ion radicals may be predominant. However, in nonpolar solvents, the hydrogen atom transfer process becomes predominant because solvation to give the separated ion radicals is suppressed. The charge-transfer mechanism was also invoked by Roth and Lamola (27) for $p,p'$-disubstituted benzophenone-1,4-diaza-bicyclo[2.2.2]octane(DABCO) systems to explain the results of investigations by means of CIDNP measurements.

In contrast to the case of carbonyl compounds, there are rather few such studies concerning the hydrogen abstraction reaction of excited aromatic hydrocarbons. In particular, there has been no attempt to do more-or-less direct measurements on the hydrogen abstraction process of excited hydrocarbons by laser flash photolysis. Assuming reaction mechanisms [19] and [20], we have tried to observe the transient radicals produced by laser flash photolysis. The results of measurements for the pyrene-aniline-$n$-hexane system are indicated in figure 5. We can observe absorption bands in the 370 - 440 nm region which are similar to the T-T absorption bands of pyrene. The characteristic 415 nm peak was already present at 50 nsec after the pulsing and its intensity was almost constant during 200 nsec. In addition to the absorption bands in the 370 - 440 nm region, we have observed weak absorption bands in the 450 - 500 nm region, whose assignment is not very clear at present. It might be due to some unknown reaction intermediate. The mechanism of intersystem crossing or triplet formation in excited charge-transfer systems is under lively investigation at present (28). There is apparent controversy concerning the mode of triplet-state generation, $i.e.$ the fast prefluorescence or preionic dissociation intersystem-crossing mechanism (5, 29) and the slow intersystem crossing matching the decay of the exciplex fluorescence or the recombination of the dissociated ion radicals (30, 31). As far as the present result in n-hexane solution is concerned, the fast intersystem crossing mechanism seems to be favored.

The absorption bands in the 370 - 440 nm region are not only due to the T-T absorption but also appear to involve the contribution from radicals produced by hydrogen abstraction of excited pyrene. By subtracting the T-T absorption bands from the observed transient spectra, we have obtained the absorption bands at *ca.* 400 nm, as indicated in figure 5,

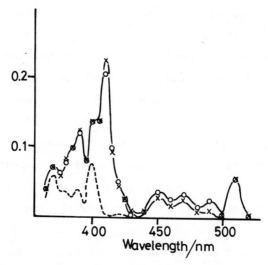

*Fig. 5 Transient absorption spectra of the pyrene-anniline-n-hexane system. The ordinate is the absorbance. Delay time from the laser pulse, o: 100 nsec.                     X: 200 nsec.*
*----: spectrum obtained by subtracting the T-T absorption band from the observed absorption band in the wavelength range of 430 - 370 nm.*

which might be ascribed to the neutral radical. Spectra similar to the one we are assigning to the neutral radical were observed more clearly in the transient spectra obtained by the conventional flash photolysis studies on the pyrene-aniline-cyclohexane system. At delay time of 50 µsec, the T-T absorption band becomes much weaker while the radical band remains fairly strong (19). On the other hand, by photolyzing 1,2-dihydropyrene in a rigid matrix at $77^{O}$K, we observed an absorption spectrum at about 400 nm, which may be ascribed to the photoproduct radical and is rather similar to that indicated in figure 5. With regard to these results, we made a semi-empirical open-shell SCFMO calculation on the π-electronic state of the odd alternant hydrocarbon. The result of calculation indicates strong absorption bands at 3.7 eV and 4.9 eV, respectively, of which the lower energy one might be identified with the spectrum observed by means of the low-temperature photolysis of 1,2-dihydropyrene and by laser flash photolysis as well as conventional flash photolysis of the pyrene-aniline system. Thus, it seems to be established that the pyrene triplet state as well as radicals due to hydrogen atom transfer from aniline to pyrene are transiently formed in the excited state of the pyrene-aniline-hexane system. However, the amount of these transients formed in the photolysis seems to be rather small indicating that a large fraction of the encounter collision complexes is converted to the ground state by a rapid radiationless process. The fact that in the quenching of excited pyrene by aniline only a small amount of triplet state is produced was recognized also by Thomas *et al.* (32). Nevertheless, since we have confirmed (19) that pyrene fluorescence is completely quenched and no exciplex fluorescence appears in the case of the solutions examined by the laser flash photolysis , the pyrene triplet state and radicals seem to be produced very rapidly from the encounter charge-transfer complex.

In acetonitrile solution, the result of laser flash photolysis of the pyrene-aniline system is quite different from that in *n*-hexane solution. In addition to the absorption in the 370 - 420 nm region, the transient spectra show strong

absorption bands of radical ions. By subtracting the
absorption bands of the pyrene anion and DMA cation from the
transient spectra at 50 nsec after pulsing, we have obtained
absorption bands which are rather similar to the T-T spectrum.
Whether this apparent T-T spectrum contains a contribution
from the neutral radical, as in the case for the $n$-hexane
solution, is not very clear. The growing-in of the triplet
state due to the recombination of the radical ions was also
observed.

From the above results, the reaction mechanism may be
summarized as follows. In nonpolar solvents, the charge-
transfer interaction in the encounter complex leads to the
rapid hydrogen atom transfer, conversion to the ground state,
as well as the triplet-state formation. That is, a large
fraction of the radical pairs A-H . . . D will go back very
rapidly to the ground state and only a small fraction will
dissociate into Ȧ-H+Ḋ. At the same time, the pyrene triplet
state may be produced rapidly, in competition with dissociation and conversion to the ground state. Contrary to this,
in strongly polar solvents, charge transfer in the encounter
collision is followed by a rapid solvation into the dissocia-
tions.

IONIC DISSOCIATION MECHANISM OF EXCITED CHARGE TRANSFER

SYSTEMS

Ionic dissociation is one of the most fundamental and
important processes of the exciplex systems. By means of
laser flash photolysis and transient photocurrent measurements, we have made more-or-less detailed studies on the
ionic dissociation processes of some exciplex systems (33)
and excited TCMB complexes (4). In the case of the

TCNB-benzene complex in 1,2-dichloroethane, it was found that the ionic dissociation does not occur *via* the relaxed fluorescent state but occurs from the excited Franck-Condon state (4). For the pyrene-DMA system, it has been shown that the free ions are formed not through the relaxed exciplex states but rapidly from the nonrelaxed charge-transfer state (33). However, in the case of the pyrene-*p*-dicyanobenzene (DCNB) system in dichloromethane, we cannot observe the rapid rise curve of the photocurrent which shows the dissociation from the nonrelaxed state. It takes about 200 - 300 nsec to achieve the peak photocurrent. This rise time is too long to regard the dissociation as occurring even from the relaxed fluorescent exciplex state. Presumably, the solvated ion-pair state may be formed rapidly at first, and free ions may be formed slowly from this relaxed ion-pair state. Contrary to this, in the case of the pyrene-NMA-dichloromethane system, we can observe both rapid and slow components in the photocurrent rise curve. Also, in this case, the rise time of the slow component is rather long (~300 nsec). Therefore, the free-ion formation in the case of the pyrene-DMA-dichloromethane system seems to occur both from the nonrelaxed state and the relaxed ion-pair state. Thus, these two systems show quite different behavior in free-ion formation, in spite of the fact that the free energies of the dissociated ion-radical state, $F(A_s^- + D_s^+)$, of these two systems are practically equal to each other in various solvents of different polarities. This result seems to indicate that the difference in chemical structure of the quenchers subtly affects the ionic dissociation process in the encounter collision.

According to our recent studies, the chemical structure of the quencher is reflected in the ionic photodissociation yield of pyrene-acceptor systems in polar solvents measured by the transient photocurrent (34). We have used 13 acceptor quenchers including quinone, nitriles, anhydrides and phthalates, in acetone and acetonitrile solutions. Roughly speaking, in the same chemical group, e.g. among the nitriles, the ionic dissociation yield decreases with the increase of the electron affinity of the acceptor; namely, for DCNB, TCNB and tetracyanoethylene(TCNE) in acetonitrile, the relative dissociation quantum yields are 1.00, 0.50 and 0.32, respectively. The same tendency was reported previously by us (4) for the ionic photodissociation of TCNB-methyl-substituted benzene complexes. That is, as the ionization potential of the donor decreases and the degree of charge transfer in the excited Franck-Condon state increases, the ionic dissociation yield decreases. This result was explained as being due to the faster nonradiative direct degradation to the ground state in the stronger complex which competes with ionic dissociation. The same reasoning may be valid for the above pyrene-cyano compound systems. However, this is valid only within the same chemical group and no definite relation between the yield and electron affinity of acceptor quencher can be established which covers different chemical groups. Since the free energy difference between the fluorescent state of pyrene and the relaxed ion pair state is larger than 10 kcal/mole for all of 13 acceptor systems, the quenching reaction may be a diffusion controlled electron transfer reaction according to the criterion of Rehm and Weller (20). The dependence of the dissociation yield upon the chemical structure of the acceptor seems to indicate that the chemical structure subtly affects the radiationless processes (other than the ionic dissociation) in the ion-pair-like complex formed after the electron transfer.

The dependence of the ionic dissociation yield of the

excited charge-transfer system on the solvent polarity is one of the most fundamental problems of exciplex chemistry. Generally speaking, the dissociation yield increases with the increase of the solvent polarity (4, 33), which has been understood qualitatively on the bases of the calculations of the free energies of the ion pairs and dissociated ion radical states (33).

Recently, we have established an empirical relation between the ionic dissociation yield and the dielectric constant of the solvent (35). Namely, equation [21] has been confirmed to hold for TCNB-toluene and pyrene-DMA systems in various solvents (35).

$$\log\left(\frac{1}{\phi} - 1\right) = \frac{p}{\varepsilon} + q \qquad [21]$$

where $\phi$ is the ionic dissociation yield, $\varepsilon$ is the static dielectric constant of the solvent and p and q are constants independent of the solvent polarity. As an example, the result for pyrene-DMA system is shown in figure 6. Since $\phi$ is considerably smaller than unity, $\log[(1/\phi)-1]$ is rather close to $\log(1/\phi)$. Therefore, equation [22] also holds approximately.

$$\log\left(\frac{1}{\phi}\right) = \frac{p'}{\varepsilon} + q' \qquad [22]$$

We can derive equations [21] and [22] by assuming appropriate models (35). One is Onsager's ion recombination model (36) which has been used in radiation chemical problems. According to this model, it is assumed that the fraction $\alpha$ of the dissociative state in the encounter collision goes to the ion-pair state with distance $r_o$, and that the dissociation and recombination of the pair compete with each other. The escape of the ions from the influence of the mutual coulomb field determines the ionic dissociation yield which may be given by

137

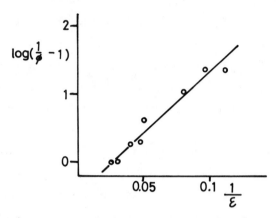

*Fig. 6 Log[(1/φ) - 1] vs 1/ε for the pyrene-DMA system.*

$$\phi = \alpha \exp\left(- \frac{r_c}{r_o}\right) \qquad [23]$$

where $r_c = (e^2/\varepsilon kT)$. From equation [23], we obtain

$$\log\left(\frac{1}{\phi}\right) = -\log\alpha + \left(\frac{2}{2.303\ r_o kT}\right)\left(\frac{1}{\varepsilon}\right) \qquad [24]$$

which is the same form as equation [22]. Another treatment is based on Horiuchi-Polanyi's relation for the activation energy (37). We assume that the dissociation quantum yield is given by

$$\phi = \frac{k_i}{k_i + k_q} \qquad [25]$$

where $k_i$ is the rate constant of the ionization from the dissociating state, $k_q$ is the rate constant of the radiationless processes other than the ionic dissociation and $k_q$ is assumed to be independent of the solvent polarity. We put

$$k_i = Z \exp\left(- \frac{E}{RT}\right) \qquad [26]$$

where the activation energy E may depend upon the solvent

polarity. The dependence of E upon the solvent polarity may be related to that of the energies of the final state of ionic dissociation by means of Horiuchi-Polanyi's relation. Since the free energy of the ion radical in a solvent of dielectric constant may be written as $F = (P/\varepsilon) + Q$ (where P and Q are constant independent of $\varepsilon$), the change of F ($\Delta F$) with solvent polarity is proportional to $(1/\varepsilon)$. The change of E ($\Delta E$) with solvent polarity is proportional to $\Delta F$ according to the Horiuchi-Polanyi relation

$$\Delta E = \alpha \, \Delta F \qquad [27]$$

Therefore,

$$k_i = const \cdot \exp\left(-\frac{\alpha \Delta F}{RT}\right) \qquad [28]$$

By means of equations [25] - [28], one can easily derive equation [29].

$$\log\left(\frac{1}{\phi} - 1\right) = const' + \left(\frac{\alpha P}{2.303RT}\right) \left(\frac{1}{\varepsilon}\right) \qquad [29]$$

A PRELIMINARY RESULT OF THE PICOSECOND LASER FLASH PHOTOLYSIS STUDY ON THE ABSORPTION SPECTRA OF THE ANTHRACENE-DEA EXCIPLEX

As described above, studies on the electronic spectra and dynamic behaviors of excited EDA systems have been made mainly with the nsec laser flash photolysis method. However, in view of the very rapid processes associated with exciplex systems, investigations by means of psec laser flash photolysis may be required and seem to be important for the elucidation of the behavior of exciplexes.

We are now making some psec studies as already described briefly for the investigation of excited dianthracene. We are using for excitation the second harmonic of a single pulse taken out from a mode-locked ruby laser. The psec continuum light for monitoring the spectra is obtained by focusing the pulse into various materials. According to our measurements, the spectrum of the anthracene-DEA-$n$-hexane exciplex system ([anthracene] = $2 \times 10^{-3}$ $M$, [DEA] = 0.7 $M$), observed 500 psec after the exciting pulse, shows a quite broad absorption band in the wavelength region of 400 - 630 nm, which is similar to the superposition of absorption bands of the anthracene anion and the DEA cation (38). The spectrum seems to be somewhat different from that observed by the nsec laser flash photolysis method (25). However, more detailed studies are necessary before any final conclusion is possible.

ACKNOWLEDGEMENT

    This paper is a summary of recent work done in collaboration with the following individuals:  T. Okada, H. Masuhara, N. Nakashima, S. Masaki, T. Hayashi, T. Hino, Y. Kume and M. Murakawa of our laboratory, and S. Misumi and Y. Sakata of the Institute of Scientific and Industrial Research of Osaka University.

REFERENCES

1. N. Mataga and Y. Murata, *J. Amer. Chem. Soc.* $\underline{91}$, 3144 (1969): T. Kobayashi, K. Yoshihara and S. Nagakura, *Bull. Chem. Soc. Japan* $\underline{44}$, 2603 (1971); Y. Torihashi, Y. Furutani, K. Yagii, N. Mataga and A. Sawaoka, *ibid.* $\underline{44}$, 2985 (1971).

2. K. Egawa, N. Nakashima, N. Mataga and C. Yamanaka, *ibid.* $\underline{44}$, 3287 (1971).

3. R. Potashnik and M. Ottolenghi, *Chem. Phys. Letters* $\underline{6}$, 525 (1970); H. Masuhara and N. Mataga, *ibid.* $\underline{6}$, 608 (1970); *Z. Phys. Chem., N.F.* $\underline{80}$, 113 (1972); H. Masuhara, N. Tsujino and N. Mataga, *Bull. Chem. Soc. Japan* $\underline{46}$, 1088 (1973); N. Tsujino, H. Masuhara and N. Mataga, *Chem. Phys. Letters* $\underline{21}$, 301 (1973); H. Masuhara and N. Mataga, *ibid.* $\underline{22}$, 305 (1973).

4. H. Masuhara, M. Shimada and N. Mataga, *Bull. Chem. Soc. Japan* $\underline{43}$, 3316 (1970); H. Masuhara, M. Shimada, N. Tsujino and N. Mataga, *ibid.* $\underline{44}$, 3310 (1971); M. Shimada, H. Masuhara and N. Mataga, *ibid.* $\underline{46}$, 1903 (1973); M. Irie, H. Masuhara, K. Hayashi and N. Mataga, *J. Phys. Chem.* $\underline{78}$, 341 (1974).

5. H. Masuhara, N. Tsujino and N. Mataga, *Chem. Phys. Letters* $\underline{12}$, 481 (1972); N. Tsujino, H. Masuhara and N. Mataga, *ibid.* $\underline{15}$, 357 (1972).

6. See for example: Th. Förster, *Angew. Chem. Int. Ed.* $\underline{8}$, 333 (1969); A. Weller, *Pure Appl. Chem.* $\underline{16}$, 115 (1968); J.B. Birks, "Photophysics of Aromatic Molecules." Wiley-Interscience, New York, 1970; N. Mataga and T. Kubota, "Molecular Interactions and Electronic Spectra." Marcel Dekker, Inc., New York, 1970; M. Ottolenghi, *Accounts*

*Chem. Res.* <u>6</u>, 153 (1973).

7.  H. Knibbe, K. Röllig, F.P. Schäfer and A. Weller, *J. Chem. Phys.* <u>47</u>, 1184 (1967).

8.  N. Mataga, T. Okada and N. Yamamoto, *Chem. Phys. Letters,* <u>1</u>, 119 (1967).

9.  T. Okada, T. Fujita, M. Kubota, S. Masaki, N. Mataga, R. Ide, Y. Sakata and S. Misumi, *Chem. Phys. Letters* <u>14</u>, 563 (1972); R. Ide, Y. Sakata, S. Misumi, T. Okada and N. Mataga, *Chem. Commun. 1009 (1972); T. Okada,* N. Mataga, Y. Sakata and S. Misumi *et al.,* to be published.

10. Complete experimental results and discussions will be published elsewhere by S. Masaki, N. Mataga, T. Okada, Y. Sakata and S. Misumi.

11. F. Schneider and E. Lippert, *Ber. Bunsenges. Phys. Chem.* <u>72</u>, 1155 (1968).

12. H. Beens and A. Weller, *Chem. Phys. Letters* <u>3</u>, 666 (1969).

13. F. Schneider and E. Lippert, *Ber. Bunsenges. Phys. Chem.* <u>74</u>, 624 (1970).

14. Details will be published by N. Nakashima, M. Murakawa and N. Mataga.

15. T. Hayashi, N. Mataga, Y. Sakata and S. Misumi, to be published.

16. H. Beens, Thesis, Free University, Amsterdam *(1969).*

17. N. Mataga, T. Okada and H. Oohari, *Bull. Chem. Soc. Japan* <u>39</u>, 2563 (1966).

18. N. Mataga and K. Ezumi, *ibid.* <u>40</u>, 1355 (1967).

19. Details will be published by T. Okada and N. Mataga.

20. D. Rehm and A. Weller, *Israel J. Chem.* <u>8</u>, 259 (1970); T. Okada, H. Oohari and N. Mataga, *Bull. Chem. Soc. Japan* <u>43</u>, 2750 (1970); K. Yoshihara, T. Kasuya, A. Inoue and S. Nagakura, *Chem. Phys. Letters* <u>9</u>, 469 (1971);

N. Nakashima, N. Mataga and C. Yamanaka, *Int. J. Chem. Kinetics* 5, 833 (1973).

21. N.C. Yang and J. Libman, *J. Amer. Chem. Soc.* 95, 5783 (1973).

22. N.J. Turro, "Molecular Photochemistry." W.A. Benjamin, Inc., New York, 1967.

23. S.G. Cohen, A. Parola and G.H. Parsons, *Chem. Rev.* 73 , 141 (1973); J.B. Guttenplan and S.G. Cohen, *J. Amer. Chem. Soc.* 94, 4040 (1972).

24. P.J. Wagner and R.A. Leavitt, *ibid.* 95, 3669 (1973).

25. S. Arimitsu and H. Masuhara, *Chem. Phys. Letters* 22, 543 (1973).

26. S. Arimitsu, H. Masuhara, N. Mataga and H. Tsubomura, to be published in *J. Phys. Chem.*

27. H.D. Roth and A.A. Lamola, *J. Amer. Chem. Soc.* 96, 6270 (1974).

28. M. Ottolenghi, *Accounts Chem. Res.* 6, 153 (1973).

29. C.R. Goldschmidt, R. Potashnik and M. Ottolenghi, *J. Phys. Chem.* 75, 1025 (1971); N. Orbach, R. Potashnik and M. Ottolenghi, *ibid.* 76, 1133 (1972).

30. E.J. Land, J.T. Richards and J.K. Thomas, *ibid.* 76, 3805 (1972).

31. H. Schomburg, H. Staerk and A. Weller, *Chem. Phys. Letters* 21, 433 (1973); *ibid.* 22, 1 (1973).

32. M.B. Cooper, R. Cooper and J.K. Thomas, to be published.

33. Y. Taniguchi, Y. Nishina and N. Mataga, *Bull. Chem. Soc. Japan* 45, 764 (1972); Y. Taniguchi and N. Mataga, *Chem. Phys. Letters* 13, 596 (1972); Y. Taniguchi, Y. Nishina and N. Mataga, *Bull. Chem. Soc. Japan* 46, 1646 (1973).

34. H. Masuhara, H. Akazawa and N. Mataga, to be published in *Bull. Chem. Soc. Japan.*

35. Details will be published by H. Masuhara, T. Hino and N. Mataga, in *J. Phys. Chem.*

36. See for example, A. Mozumder *in* "Advances in Radiation Chemistry." (M. Burton and J.L. Magee, eds.), Vol. 1, p. 83, Wiley-Interscience, New York, 1969.

37. S. Glasstone, K.J. Laidler and H. Eyring, "The Theory of Rate Processes." p. 145. McGraw-Hill, New York, 1941.

38. Y. Kume, N. Nakashima and N. Mataga, to be published.

# HYDROGEN EXCIMER - A BOUND STATE OF $H_4$

JOSEF MICHL* AND RONALD D. POSHUSTA**

*Department of Chemistry, University of Utah
Salt Lake City, Utah 84112

**Chemical Physics Program, Washington State
University, Pullman, Washington 99163

## INTRODUCTION

Simple quantum mechanical arguments lead one to expect that nonpolar closed shell molecules are not likely to form significantly bound dimers in their ground state, but are likely to form such dimers in their excited states. The simplest known examples of such excited dimers are the bound excited states of molecular helium, $He_2$*, which might be considered to represent "helium atomic excimers". The simplest molecular excimer should be formed from a ground-state hydrogen molecule and an excited-state hydrogen molecule and have the formula $H_4$*. No such "hydrogen molecular excimer" has ever been observed to our knowledge. Nevertheless, such a species might be of considerable interest for a variety of reasons, and we presently predict its (short-lived) existence and calculate some of its properties.

A prime reason for interest in hydrogen excimer is its very simplicity. High quality *ab initio* calculations at a variety of geometries should be financially feasible and might provide some insight into the nature of bonding in excimers in general. Further, a knowledge of the potential energy hypersurfaces of the $H_4$ system might permit predictions of photochemical reactions of molecular hydrogen and of the ion recombination process, $H_3^+ + H^-$, and detailed knowledge of the theoretical aspects of the behavior of such simple systems might be helpful for an understanding of photochemical and ion-recombination reactions in general.

Previous calculations of the $H_4$ system (1-6) concentrated mainly on the repulsive ground state. Results for excited states at a few selected geometries have also appeared (3-6). Here we report results of the first stage of our calculations, involving a study of the lowest three singlet states of $H_4$ in a limited region of its six-dimensional nuclear configuration space by a relatively simple *ab initio* method, and concentrate on the properties of the bound state of $H_4$ which was thus discovered. Since this state correlates with a ground state $H_2(X^1\Sigma_g^+)$ and a singlet excited state $H_2^*(B^1\Sigma_u^+)$ in the dissociation limit, it can indeed be appropriately called hydrogen excimer.

METHOD OF CALCULATION

Our calculations were performed in the VB formalism (7) using a program which had previously proved valuable (8) for molecules of type $H_n^+$. Since all VB structures are included, the calculation is equivalent to a full configuration interaction calculation. A minimum basis set of floating 1s orbitals was used; these were contractions (9) of four GTO's. The exponent of each atomic orbital was optimized separately for each state and each geometry. Orbitals were also separately polarized by floating the centers of each member of a contraction (10).

The advantage of the use of a minimum basis set is low
cost, which permits rapid mapping of large areas of the six-
dimensional space, and the easy intuitive understanding of the
nature of the resulting wavefunctions in terms familiar to
chemists and possibly transferable to other molecules. There
are three commonly recognized disadvantages. The first is
the inability to allow for atomic polarization in the wavefunction. However, in our calculation, the same purpose is
achieved by floating the atomic orbitals to optimum positions.
The second and more serious disadvantage is the inability to
describe intraatomic electron correlation. Thus, $H^-$ is calculated not to be bound, and all states in whose VB description structures with $H^-$ have large weight are probably
described relatively poorly with respect to covalent states.
For instance, the $B^1\Sigma_u^+$ state of $H_2$, which is well represented by $H^+H^- \longleftrightarrow H^-H^+$, is calculated to have a minimum near
1.9 $\overset{\circ}{A}$ with energy -0.6795 a.u., while the exact calculation
of Kolos and Wolniewicz (11) yields 1.28 $\overset{\circ}{A}$ and -0.7566 a.u.
This can be contrasted with results for the largely covalent
$X^1\Sigma_g^+$ state of $H_2$, where floating minimum basis set results
(0.74 $\overset{\circ}{A}$, -1.1502 a.u.) are quite close to the exact ones (11)
(0.74 $\overset{\circ}{A}$, -1.17447 a.u.). Also the comparison of results of
Wilson and Goddard (3) with those of Rubinstein and Shavitt
(6) is instructive in this regard. We will have to keep this
in mind when comparing energies of states which differ
greatly in the weight of ionic structures in their wavefunctions.

The third main disadvantage is the inability to describe
more than the few lowest states of $H_4$, and to provide the
proper dissociation limits for most states, even in the complete valence CI approach used. Neither H nor $H^-$ have any

excited states in this approximation. In $H_2$, $X^1\Sigma_g^+$ and $b^3\Sigma_u^+$ states are represented reasonably, but $B^1\Sigma_u^+$ and $E^1\Sigma_g^+$ states are represented as hybrids of $H^+H^-$ and $H^-H^+$ at all distances, whereas correctly they should dissociate to $H + H^*$. Moreover, higher states such as the $C^1\Pi_u$ and $a^3\Sigma_g^+$ states are missing altogether.

In defense of our use of the floating minimum basis set, it should be pointed out that if excited states of other polyatomics are any guide, it seems certain that at geometries at which the four nuclei are relatively close together, the first few low-lying states will be "valence" states and that their description, including relative diffuseness, will not be totally unreasonable considering that we optimize exponents separately for each state. After all, at moderate internuclear separations, the lowest three states of $H_2$ are qualitatively properly described in this basis set, and low-lying states of $H_4$ at relatively short bond lengths will be the only ones of interest here.

Finally, in order to secure a firmer basis for interpretations based on minimum basis set results, we have repeated calculations for selected geometries with 1s1s' and 1s2s basis sets, which at least partially removes many of the above objections. For instance, with a 1s1s' basis, the $B^1\Sigma_u^+$ state of $H_2$ has a minimum at 1.6 Å with energy -0.70876 a.u., and dissociates into uncharged species.

RESULTS AND DISCUSSION

*HYDROGEN EXCIMER*

The low-lying states of $H_2$ which will be of interest to us are the $X^1\Sigma_u^+$ ground state (both electrons in the bonding sigma MO), and the singlet $B^1\Sigma_u^+$ and triplet $b^3\Sigma_u^+$ singly excited states (one electron in the bonding sigma MO, the other in the antibonding sigma MO).

Formally, low-lying valence singlet states of $H_4$ can be expected to result from an approach of two $H_2$ molecules in their $X^1\Sigma_g^+$ ground state, or one $H_2$ in $X^1\Sigma_g^+$ ground and the other in $B^1\Sigma_u^+$ excited state, or two $H_2$ molecules in their $b^3\Sigma_u^+$ triplet states coupled into an overall singlet. Only the second of these approaches can yield what would commonly be called an excimer and indeed, the potential energy hyper-surface of this approach is the only one of the three for which we have been able to find a bound minimum so far. Depending on the nuclear geometry, this "excimer" surface is second or third in energy among the singlet states of $H_4$.

At the present level of our calculation, the minimum geometry for the excimer is kite-shaped, close to a square with sides of 1.25 Å. It can be reached in a reaction without activation energy by combining a ground-state $H_2(X^1\Sigma_g^+)$ with an excited $H_2(B^1\Sigma_u^+)$. The process is calculated to be exothermic by about 25 kcal/mole. No state crossing or avoided crossings are in the way of a planar trapezoidal or rectangular approach and the reaction

$$H_2(X^1\Sigma_g^+) + H_2(B^1\Sigma_u^+) \rightleftharpoons H_4^* \text{ (excimer)}$$

can be expected to proceed adiabatically.

The 1.25 × 1.25 Å square form of the excimer is calcu-lated to lie at lower energy than any rectangle and is also stable to all other distortions except for an $e_u$ mode. This degenerate mode converts the square into a trapezoid or a kite (figure 1). The best trapezoid has parallel sides of lengths 0.95 Å and 1.6 Å located 1.35 Å apart and is 4 kcal/mole below the square. The best kite has diagonals 2.8 Å and 1.85 Å long, with a short side of 1.05 Å and a long side of 1.6 Å; its energy is 11 kcal/mole below that of the square.

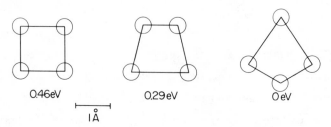

0.46eV          0.29eV          0 eV

$$\vdash\!\!-\!\!\overset{\circ}{1\,\mathrm{A}}\!\!-\!\!\dashv$$

*Fig. 1   Best square, best trapezoid and best kite geometries of the $H_4^*$ excimer.*

The six normal modes of vibration for the kite-shaped molecule are shown in figure 2.

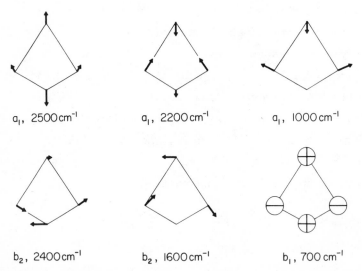

$a_1$, 2500 cm$^{-1}$          $a_1$, 2200 cm$^{-1}$          $a_1$, 1000 cm$^{-1}$

$b_2$, 2400 cm$^{-1}$          $b_2$, 1600 cm$^{-1}$          $b_1$, 700 cm$^{-1}$

*Fig. 2   Normal vibration modes of the $H_4^*$ excimer.*

Three have $a_1$ symmetry and their wave numbers are at present calculated to be 1000, 2200, and 2500 cm$^{-1}$. Two have $b_2$ symmetry and wave numbers 1600 and 2400 cm$^{-1}$. The wave number of the out-of-plane bending $b_1$ mode is 700 cm$^{-1}$.

This first-order vibrational analysis may not be completely meaningful since the hydrogen nuclei are very light and the barrier separating one kite geometry from another seemingly rotated by about $90^{\circ}$ is small, comparable in size to the inversion barrier in ammonia. Quite small changes in nuclear positions suffice to convert one such kite into another (figure 3). The best path for this pseudorotation avoids the square geometry and proceeds through the best trapezoid. Thus, the $H_4$ excimer can be expected to show a

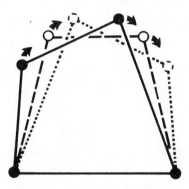

*Fig. 3  Pseudorotation in $H_4^*$ excimer.*

"pseudorotational level quadrupling" similarly as ammonia exhibits inversion doubling. We hope to calculate the level splittings and predict possible microwave absorption wavelengths after completing extended basis set calculations for this region of the excimer hypersurface.

At the square ($D_{4h}$) geometry, the symmetry of the electronic wavefunction of the excimer is $^1B_{2g}$; the ground state has $^1B_{1g}$ symmetry and the state derived from two $H_2$ molecules in their triplet states is of $^1A_{1g}$ symmetry. Thus, electric dipole radiative transitions between the three states are forbidden. When the symmetry is lowered to a kite or a trapezoid, the symmetry labels change appropriately and the transitions become symmetry-allowed.

The best simple representation for the nature of the

wavefunction of the excimer is a combination of $H_3^+$ and $H^-$:

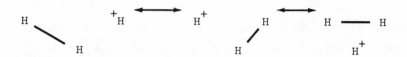

These VB structures clearly indicate the parentage of the species to be $H_2(X^1\Sigma_g^+)$ and $H_2(B^1\Sigma_u^+)$ and show that both excitation exchange and charge transfer contributions to the structure are important.

The excimer can also be viewed as derived from the $H_2(B^1\Sigma_u^+)$ state, which can be described as $H^+H^- \longleftrightarrow H^-H^+$, by replacing $H^+$ by the well-known stable entity $H_3^+$. This immediately suggests that the $H_4^*$ excimer might also be formed in another process, namely the ion recombination reaction

$$H_3^+ + H^- \rightleftharpoons H_4^* \text{ (excimer)}.$$

Work in progress indicates that Born-Oppenheimer surfaces indeed connect properly for such an adiabatic process, but an avoided surface crossing which occurs along this reaction path, although avoided fairly strongly, might cause radiationless transitions to lower surfaces and thus thwart the (endothermic) formation of ions from $H_4^*$ excimer and the (exothermic) ion recombination to give the excited $H_4^*$ species. If the latter process occurs nevertheless, it will be very interesting since excited $H_2(B^1\Sigma_u^+)$ could be the end product under suitable conditions. This process would then represent one of the simplest chemiluminescent reactions and might possibly lead to the first vacuum uv chemical laser, perhaps using suitable isotopes or using other simple negative ions in place of $H^-$. Current interest in laser fusion makes it quite appealing to pursue this line of inquiry.

At its minimum geometry, the excimer lies about 24 kcal/ mole above the lowest excited state (which originates in singlet coupling of two triplet $H_2$'s) and about 120 kcal/mole above the ground state. This ordering of states is preserved at the best trapezoid and best square geometries and in a fairly large neighborhood of these geometries as well. In general, a decrease in the size of the $H_4$ species favors the excimer state relative to the state which originates in the singlet coupling of two triplet $H_2$ molecules. Eventually, they cross and the excimer state then represents the lowest excited state of the molecule. This behavior is qualitatively reasonable since the excimer state is purely ionic whereas the state originating in two triplet $H_2$'s contains both ionic and covalent contributions. In MO language, the latter state can be described as a doubly excited state and we shall use this convenient label in the following discussion. The relative weight of the covalent contributions in this state increases as the size of the molecule grows and, in the limit of four separated hydrogen atoms, the doubly excited state is purely covalent and degenerate with the ground state, from which it differs only in the spin coupling scheme.

Because of the previously described deficiencies of a minimum basis set, the calculated energy of the ionic excimer state undoubtedly contains a larger error than that of the less ionic doubly excited state [which in turn is more ionic than the ground state - $cf.$ Porter and Raff (4)], but this differential error becomes smaller as the size of the $H_4$ species decreases. The 24 kcal/mole energy difference mentioned above is therefore likely to represent an upper limit, and the states may lie closer together. However, double zeta

basis set calculations of Rubinstein and Shavitt (6) for
square geometries gave the same order of states as ours. We
have performed a similar calculation at the "best kite" geometry and again found the same order of states, with the excimer state now only a little more than 1 kcal/mole above the
lowest excited state. Since at this level of approximation
the minimum in the excimer surface may lie at a different
geometry, we cannot exclude the possibility that the excimer
state is the lowest excited state at its equilibrium geometry.
We hope to clarify this point by future calculations using
larger basis sets.

In principle, the $H_4^*$ excimer might be detectable, for
instance, by its emission to the ground state (continuum near
2400 $\overset{o}{A}$), by its transient absorption (wavelengths of which
we cannot predict from calculations performed so far), and
perhaps even by its pseudorotational microwave absorption
spectrum if its lifetime is sufficiently long. However, the
presence of a lower-lying state separated by only a small
energy gap leads one to suspect that rapid radiationless
deactivation may compete efficiently with radiative decay and
might shorten the lifetime of the species. Also, crossings
with triplet hypersurfaces (not calculated) might intervene.

In both possible modes of adiabatic formation, $H_2(X^1\Sigma_g^+)+$
$H_2(B^1\Sigma_u^+)$ and $H_3^+ + H^-$, vibrational and/or rotational energy
removal by a third body will be required in order to stabilize the $H_4^*$ excimer. Even in the absence of such stabilization, information about the potential energy hypersurface
might be accessible, e.g., from ion scattering experiments
on $H_3^+ + H^-$, using emission by the $H_2(B^1\Sigma_u^+)$ product for
detection.

The fate of the four hydrogen atoms of the excimer after
radiative or radiationless conversion to one of the two lower

states is of great interest with regard to the photochemistry of molecular hydrogen. Our search has already revealed an amazing variety of possible reaction paths, several of them with analogies in organic photochemistry, but lack of time prevents any detailed discussion here.

*RELATIONS TO LARGER MOLECULES*

While the $H_4^*$ hydrogen excimer is of interest in itself, our work was at least partly prompted by the hope that extrapolations to larger chemical systems might be possible. Since most of the audience is primarily interested in larger molecules, it will hopefully not be entirely out of place to outline several such extrapolations, although they are necessarily highly speculative at this stage. Four possibly fruitful areas for drawing analogies have occured to us: first, the nature of bonding and structure of excimers; second, the possible relation of excimer formation to photochemical paths; third, the possible relation of excimers to ion recombination reactions; fourth, the possible relation of results on $H_4$ to the triplet-triplet annihilation process.

Of course, the absence of "core" electrons in the $H_4^*$ excimer makes it different from known organic excimers, since nothing prevents a very intimate approach of the two components. Qualitatively, one could imagine that in addition to having more than the four "optical" electrons of $H_4^*$, other excimers (*e.g.*, pyrene excimer) also have a network of sigma bonds which do not participate in optical phenomena and can be roughly approximated, say, by elastic springs which follow the Morse potential. The resulting decrease in mobility will prevent the existence of any strict analogy to phenomena such as the pseudorotation predicted for $H_4^*$. Also, the kite-shaped geometry preferred by $H_4^*$ excimer can hardly be

generalized, although it might indicate a tendency for deviations from exact parallelism of the two halves of organic
excimers (which would remove the symmetry forbiddenness of
their fluorescence). The fact that the best geometry for $H_4^*$
excimer is not square may be significant; it is in accordance
with the idea that excimers favor, or at least do not strongly
oppose, partial localization of excitation in one of the two
halves, as achieved by a suitable adjustment in geometry. The
potential energy barrier connected with transfer of excitation
from one half to the other with concommitant changes in geometry would, of course, be relatively small. Because of the
presence of the underlying core electrons which can be
expected to resist compression, stretching and other adjust-
ments to the requirements of the "optical" electrons, one can
expect this tendency to be weaker in organic excimers than in
$H_4^*$. This lowering of symmetry would provide another mechanism
making fluorescence symmetry-allowed. Finally, the highly
ionic nature of the wavefunction of $H_4^*$ excimer is in accordance with common notions concerning organic excimers.

The relation of excimer and exciplex formation to photochemical cycloaddition reactions is an area of considerable
current interest (12) and it now appears certain that at least
in some instances exciplex formation precedes product formation
It is also well known that the occurrence and regiospecificity
of numerous cycloaddition reactions can be accounted for by
simple quantum mechanical procedures which evaluate terms
related to the initial interaction of an excited molecule with
its ground state partner and thus to the stability of the excimer or exciplex (13). Yet, it is hard to believe that ground-
state products are formed directly from the exciplex. If the
ground-state potential energy hypersurface is reached at the
exciplex geometry, for instance by excimer emission, a

repulsive part of this surface is reached and the two partners diffuse apart. Nonradiative return to that region of the ground-state hypersurface which slopes downhill toward products would have to be considerably nonvertical and appears unlikely. It makes more sense to expect the existence of a driving force which causes the molecular geometry to become more like that of the product while the molecule is still excited, so that essentially vertical return to the ground-state hypersurface will reach a high-energy region from which return to the two starting molecules or collapse to products can both occur. In so far as one can extrapolate from our results for $H_4$, the excimer surface cannot be expected to provide such a driving force since only one minimum is present. Here, the doubly excited state comes into play. It will clearly have an analogy in organic $\pi$-electron systems (14) and closer examination shows that is is identical with the state known from Woodward-Hoffmann correlation diagrams which has a minimum located above the maximum of the barrier in the ground state, both the minimum and the barrier being a result of an avoided crossing for ground-state forbidden pericyclic reactions such as olefin dimerization. The importance of the minimum and of the driving force it provides for electrocyclic reactions was first recognized by van der Lugt and Oosterhoff (15), who, however, deemphasized completely the role of the absorbing state in the photochemical process (this state corresponds to our excimer state). Following a recent review article (16), we shall refer to this minimum as a "funnel" since it is likely to efficiently trap excited molecules and take them to the ground state hypersurfaces.[*]

Our present results for $H_4$ and additional unpublished results of more approximate calculations for larger molecules (17) then suggest a general picture for cycloaddition in which both excited states play a role.

During initial encounter, the nuclear motions are governed by the excimer surface, which tends to lead the molecules into close proximity corresponding to the excimer geometry, in what most likely is a reversible adiabatic process. If several orientations are possible (*e.g.*, head-to-head and head-to-tail, *syn* or *anti*), the most stable one will be preferred. If the two molecules are not identical, an analogous surface exists and should now perhaps be called exciplex surface. In many instances, it will slope similarly as above, to an exciplex minimum, but the situation is naturally more complicated than for an excimer and will require a more detailed analysis. The above mentioned simple quantum chemical arguments (13) based on interaction diagrams provide an estimate of this slope. In other instances, the interaction may be so unfavorable that no exciplex minimum exists; for example, if there is a great discrepancy in the energies of the LUMO's of the two components or of their HOMO's, or if there is too much steric hindrance. It could also happen that it exists, but cannot be reached at a given temperature because of an energy barrier such as might result from abnormal orbital crossover (18). We would like to hypothesize that the existence and accessibility of an excimer or exciplex minimum in the excimer (exciplex) state is a prerequisite for efficient singlet photocycloaddition. If the minimum does not exist, the two molecules may still reach the geometry where the minimum normally would lie, but only for short periods of time (collision complex) and other modes of deactivation are then likely to compete more successfully with the product-forming course of events described in the following, which is based on the work of van der Lugt and Oosterhoff (15) on electrocyclic reactions.

In the next step, the electronic state changes from that
of the excimer to that of the doubly excited state, corresponding to a singlet coupling of two triplet molecules.
The latter singlet state rises steeply in energy if the two
molecules move apart, and slopes down towards the funnel.
The funnel provides an efficient way to return to the ground
state, which then leads the "supermolecule" either to the
product geometry or back to the two starting species in a
ratio  which depends on details of the molecular structure,
and possibly even on the direction from which the funnel was
entered (19).

In this picture, the excimer minimum serves as a reser-
voir which keeps the molecules at a nuclear geometry suitable
for jumping over to the nearby funnel.  It is thus easy to
understand why the relative size of the population of several
possible excimer minima (head-to-tail, etc.) determines which
one of several funnels will actually be used, and thus
affects the regiospecificity of the overall reaction.  The
rate at which the funnel is reached from the various possible
excimer minima is the other important factor.  Three main
cases appear possible and are shown in figure 4.

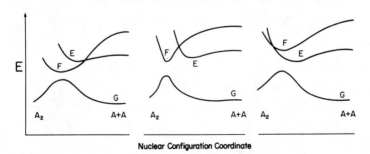

Fig. 4  Possible relations of excimer minimum
and funnel in the doubly excited state
in a photocycloaddition.

First, the state containing the funnel can lie below the minimum in the excimer state, as is the case for $H_4$. In large organic molecules, very rapid radiationless conversion would then be expected; the excimer (or exciplex) would have an extremely short lifetime and its formation would be very hard to detect by emission or transient absorption. Second, at the geometry of the minimum in the excimer state, the excimer state can be the lowest excited singlet while at the geometry of the funnel, the excimer state is the second excited singlet and the doubly excited state is lower. Then, the lowest excited singlet contains both the excimer minimum and the funnel, but the nature of its wavefunction is quite different at the two geometries. A barrier which results from (possibly avoided) state crossing then separates the excimer minimum from the funnel and its height will affect the rate at which the funnel is being reached. If the barrier is reasonably high, emission of the excimer (exciplex) may be able to compete; if the barrier is too high, emission and excimer dissociation may prevail completely over product formation. Third, the excimer state may be the lowest excited singlet both at the geometry of the excimer minimum and at the geometry of the funnel. Excimer formation should then be easy to detect, but little or no photochemical product will be formed.

The order and nature of the lowest two excited states (excimer state and doubly excited state) at geometries such as described here [ biradicaloid geometries (19) ] undoubtedly are of great importance. Simplified discussions using three-configuration CI have appeared (20, 21) but our results on $H_4$ and other biradicaloid species (17, 22) show that a more complicated description is required. This will be published elsewhere.

It is tempting to speculate that even intramolecular pericyclic reactions, such as electrocyclic reactions, proceed similarly as outlined above, *i.e.* through a special kind of intramolecular excimer, but time limitations preclude a detailed discussion here.

There are two other areas for which extrapolations from our results for $H_4$ might be of interest. One of these, ion-recombination reactions, was already briefly discussed above. Recombination of organic radical ions is, of course, well known to produce excited singlets, but such a process would correspond to a reaction $H_2^+ + H_2^-$ in our case. There appear to be no known analogies to the process $H_3^+ + H^- \rightarrow H_4^*$ (excimer) in organic chemistry, and it might be interesting to look for them.

The other area, triplet-triplet annihilation, will also be mentioned only briefly. Under random conditions, one out of nine triplet-triplet encounters will lead to an overall singlet species and in these cases the nuclear motion will be governed by the hypersurface of the doubly excited states with a funnel which has been discussed above. In the case of hydrogen, the lowest triplet $H_2$ is not a bonded species so that a discussion of an encounter of two such triplets would appear to be purely academic, but the presence of additional bonds makes triplets of organic molecules bonded and such a discussion is then of interest. *A priori,* two processes are likely to compete. The first is internal conversion from the high-energy doubly excited state in which the approach initially starts, already at the time of the initial approach, to reach the excimer hypersurface which lies at lower energies, followed by dissociation into ground state and "excimer chemistry" such as dissociation, emission, etc. This internal conversion is believed to occur quite efficiently already at large separations (23). The second is travel all the way to the funnel in the doubly excited state, followed by further "funnel chemistry", such as crossover to excimer state or return to the ground state followed by photoadduct formation or by regeneration of starting materials. Which processes are actually preferred would again depend on the detailed properties of the hypersurfaces (figure 4). While formation of ordinary excited singlets upon triplet-triplet annihilation is well known, direct photoproduct formation without prior formation of the ordinary excited singlet represents an interesting prediction of the model described here and raises the possibility that even in those instances in which direct irradiation does not lead to singlet photocycloaddition, triplet-triplet annihilation may. An intramolecular analogy of the triplet-triplet annihilation photoprocess would be a two-photon excitation into the postulated low-lying *gerade* state of polyenes (24) followed by an electrocyclic process without ever going through the well-known spectroscopically accessible *ungerade* state of the olefin. Such processes have not been described so far.

ACKNOWLEDGEMENT

The authors wish to acknowledge support from NSF (grant GP 37551X1 to J.M.) and from the Washington State University, and collaboration of Dr. Wolfgang Gerhartz on parts of this project.

REFERENCES

1. M.E. Schwartz and L.J. Schaad, *J. Chem. Phys.* *48*, 4709 (1968).

2. R.W. Patch, *J. Chem. Phys.* *59*, 6468 (1973).

3. C.W. Wilson, Jr. and W.A. Goddard III, *J. Chem. Phys.* *51*, 716 (1969).

4. R.N. Porter and L.M. Raff, *J. Chem. Phys.* *50*, 5216 (1969).

5. H. Conroy and G. Malli, *J. Chem. Phys.* *50*, 5049 (1969).

6. M. Rubinstein and I. Shavitt, *J. Chem. Phys.* *51*, 2014 (1969).

7. H. Eyring, J. Walter and. G.B. Kimball, "Quantum Chemistry." Wiley, New York, 1955.

8. R.D. Poshusta and F.A. Matsen, *J. Chem. Phys.* *47*, 4795 (1967); R.D. Poshusta and. D.F. Zetik, *J. Chem. Phys.* *58*, 118 (1973); W.I. Salmon and R.D. Poshusta, *J. Chem. Phys.* *59*, 4867 (1973).

9. S. Huzinaga, *J. Chem. Phys.* *42*, 1293 (1965).

10. J.L. Whitten, *J. Chem. Phys.* *39*, 349 (1963).

11. W. Kolos and L. Wolniewicz, *J. Chem. Phys.* *41*, 3663 (1964); *45*, 509 (1966).

12. J.B. Birks, "Photophysics of Aromatic Molecules." pp. 319,629,633, Wiley-Interscience, New York, 1970; K. Mizuno, C. Pac and H. Sakurai, *J. Amer. Chem. Soc.* *96*, 2993 (1974); R.A. Caldwell and L. Smith, *J. Amer. Chem. Soc.* *96*, 2994 (1974); O.L. Chapman and R.D. Lura, *J. Amer. Chem. Soc.* *92*, 6352 (1970); J. Saltiel, J.T. D'Agostino, O.L. Chapman and R.D. Lura, *J. Amer. Chem. Soc.* *93*, 2804 (1971); N.C. Yang, private communication.

13. W.C. Herndon, *Topics Curr. Chem.* <u>46</u>, 141 (1974).

14. J. Koutecký and J. Paldus, *Theor. Chim. Acta* <u>1</u>, 268 (1963).

15. W. Th. A.M. van der Lugt and L.T. Oosterhoff, *J. Amer. Chem. Soc.* <u>91</u>, 6042 (1969).

16. J. Michl, *Topics Curr. Chem.* <u>46</u>, 1, (1974).

17. W. Gerhartz and J. Michl, unpublished results.

18. J. Michl, *Mol. Photochem.* <u>4</u>, 287 (1972).

19. J. Michl, *Mol. Photochem.* <u>4</u>, 243 (1972).

20. J. Michl, *Mol. Photochem.* <u>4</u>, 257 (1972).

21. L. Salem and C. Rowland, *Angew. Chem., Int. Ed.* <u>11</u>, 92 (1972).

22. C.R. Flynn and J. Michl, *J. Amer. Chem. Soc.* <u>96</u>, 3280 (1974).

23. J.B. Birks, "Photophysics of Aromatic Molecules." Chap. 8 and p. 631. Wiley-Interscience, New York, 1970.

24. T.H. Dunning, Jr., R.P. Hosteny and I. Shavitt, *J. Amer. Chem. Soc.* <u>95</u>, 5067 (1973).

\* Strictly speaking, we reserve the term "funnel" for areas in which the jump occurs with such high probability that thermal equilibration in the upper state is not achieved first. This need not be the case here.

THE ACETONE EXCIMER. A CASE FOR THE CONCEPT OF

"MASKED" OR "CRYPTO" ELECTRONICALLY EXCITED STATES

NICHOLAS J. TURRO

*Chemistry Department, Columbia University*

*New York, New York 10027*

## THE IDEA AND THE OBSERVATION OF EXCIMERS

What is an excimer? A commonly accepted definition (1) is: An electronically excited molecular complex of definite stoichiometry formed from two molecules of the same species and which is dissociated in the ground state. This definition expresses an idea and not an observation and is therefore, of course, written in a *"theoretical language"*. As an addendum to the above definition, we might add the notion that an exciplex is capable of thermally dissociating into its constituent species in a manner such that one of the fragments is electronically-excited. The term excimer (equation [1]) brings to mind both that of (i) an electronically-excited particle colliding with an unexcited particle of the same type and thereby forming an associated "stoichiometric complex" and (ii) an electronically excited association which can fragment thermally into two particles which are chemically identical except that one particle is electronically excited.

$$M^* + M \;\underset{(ii)}{\overset{(i)}{\rightleftharpoons}}\; (M\text{-}\overset{*}{-}\text{-}M) \quad \textit{encounter complex} \qquad [1]$$

The ideas stimulated by equation [1] can be treated independently of certain experimental or observational details such as the method of formation of M* (or M-$\overset{*}{-}$-M) and the method of detection of these species. Also, the idea of an excimer can be considered independently of the details of a theory of bonding of M-$\overset{*}{-}$-M or of its specific mechanism of formation.

Switching now to the experimental viewpoint, a proposal of the involvement of an excimer in a system under study may rest upon one or more of the following observations (2):

(i)  concentration quenching in the determination of a photochemical parameter;

(ii) concentration dependent emission or absorption;

(iii) fitting of experimental data by a kinetic scheme which assumes the occurrence of an excimer;

(iv) formation of photochemical products which suggest an excimer precursor;

(v)  correlation of a measured photochemical parameter with a property which is theoretically expected of an excimer;

(vi) a negative temperature coefficient of some photochemical parameter;

(vii) temperature dependent emission or absorption.

Given this introduction to some of the possible conceptual and observational attitudes concerning excimers, let us now consider the specific situation for the acetone excimer.

THE ACETONE EXCIMER:  SOME THEORETICAL CONSIDERATIONS

What would we expect the properties of an acetone excimer to be?  First let us consider the situation from the

viewpoint of orbital interactions and ignore, for the time
being, the possible complication of spin.  Figure 1 depicts
the type of interactions we might expect to occur when two

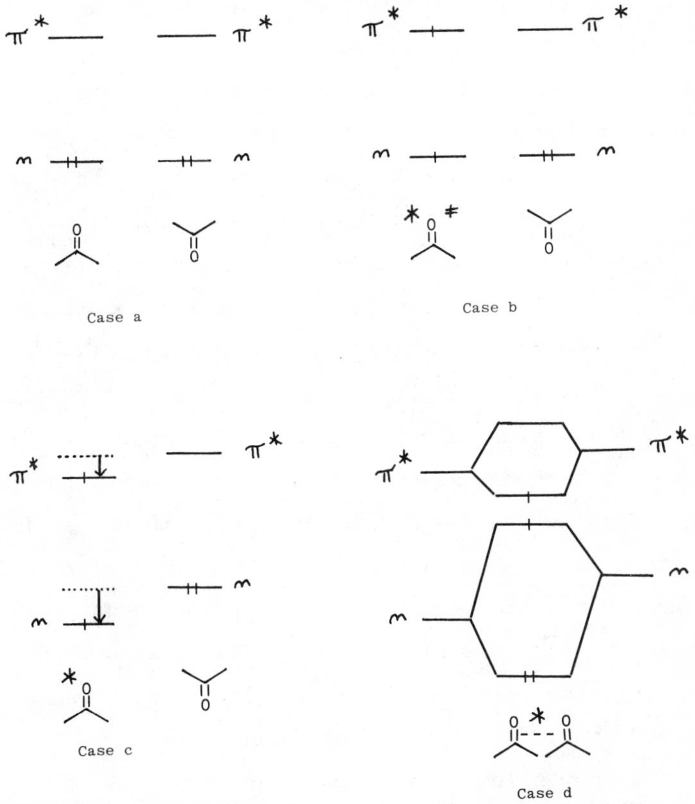

Fig. 1   Schematics of a simple orbital interaction
between an acetone $n,\pi^*$ state and a ground-
state acetone molecule.

acetone molecules come to within close approach of one another.
For two ground-state acetone molecules (case a), no significant stabilizing interactions beyond those of the van der Waals
forces or  dipole-dipole type are expected.  For the sake of
simplicity, we will consider only the highest energy occupied
(HO) orbital and the lowest energy unoccupied (LU) orbitals
of the interacting acetones.  Within this approximation we

would expect the existence of weakly bound head-to-tail dimers (or higher aggregates) of acetone. There is some experimental support for the existence of acetone aggregates (3). Upon electronic excitation of one of the two acetone molecules (case b), immediately after the n-π* excitation we have a Franck-Condon state in which the nuclear geometry of the ground-state dimer is maintained and which has the electronic configuration $(n)_L (π*)_L (n)_R^2$, where the subscript labels will pertain to the acetone on the *left* (L) or the *right* (R) of the appropriate case. This particular situation lasts only briefly, however, because thermal equilibration of the excited acetone on the left occurs rapidly. After equilibration, the energies and nuclear and electronic structures of the excited acetone change. The energies of the electron in both the n and π* orbitals (in an approximation which refers to the *new* nuclear geometry) in the equilibrated n,π* state are probably lower than the same electrons in the Franck-Condon state. Furthermore, the energy gap between the n and π* orbitals is *smaller* in the equilibrated n,π* state since the emission of ketones is strongly red-shifted ($\sim 12,000$ cm$^{-1}$) from their absorption (case c).

If we now allow an orbital interaction (4) to occur between the equilibrated n,π* acetone and a ground-state acetone, we see that a net stabilization may occur. First-order calculations indicate that the largest interaction between orbitals should occur between the n electrons on the unexcited acetone molecule and the half-filled n oribtal on the excited acetone (4b). If this is true, then the structure of the excimers should be of the head-to-head type (case d), with substantial O-O bonding.

THE ACETONE EXCIMER: SOME EXPERIMENTAL OBSERVATIONS OF

ACETONE PHOTOLUMINESCENCE AND PHOTOPHYSICS

The history of acetone luminescence in solution has had some wry twists and turns. For many years, it was believed that only weak emission maximizing at 405 nm, which was assigned to acetone *fluorescence* ($\phi_F \sim 10^{-3}$), could be achieved by photoexcitation in solution (5). However, in rigid media at 77 K, acetone photoluminescence consists mainly of *phosphorescence* ($\phi_P \sim 3 \times 10^{-2}$) (6). Thus, it was concluded that in fluid solution there is a strong quenching of acetone phosphorescence.

In 1970, strong support for the occurrence of acetone singlet excimer was reported (7). It was claimed that the oft-observed acetone fluorescence, traditionally assigned to that of monomeric acetone, was actually acetone singlet excimer emission. These results, unfortunately, were not reproducible in other laboratories (8, 9) and were subsequently retracted by the original authors (10). It thus now appears quite certain that the well-known acetone fluorescence is due to individual molecules of this ketone and not to excited aggregates.

It has been reported that the lifetime of acetone fluorescence is nearly solvent and concentration independent ($\tau_F \sim 2 \times 10^{-9}$ sec). This result requires that, *if* acetone singlet excimers are formed in significant amounts, they do not decay in a manner which affects the measured decay of monomeric excited singlet acetone. Thus, we can make the following general statements:

(i) either acetone singlet excimers $^1$(A-$\overset{*}{\text{-}}$--A) are not formed, even at high concentrations of acetone or

(ii)   acetone excimer is formed, but reversibly, and the excimer must have an energy content that is comparable to the system excited monomer $^1A*$ plus ground-state acetone, A.

With regard to photoexcited acetone phosphorescence, it was recently found that in "unreactive" and deoxygenated sol-vents,$^†$ phosphorescence is easily detected (11).  Furthermore, in "unreactive" solvents, the lifetime of acetone phosphores-cence is *not* concentration dependent.  Thus, the same two general possible conclusions obtained for acetone singlet excimer apply to the triplet excimer, *i.e.*, it is either not formed or it is in rapid equilibrium with triplet acetone, $^3A*$, and ground-state acetone, A.

In neat acetone solution (12), the rates of sensitization of *trans-cis* isomerization of 2-pentene and of biacetyl phosphorescence were found to be larger than the corresponding rates in mixed solvents of acetone and hydrocarbons, *i.e.*, the Stern-Volmer constant $k_{SENS}\tau$ was larger in neat acetone than in mixed acetone-hydrocarbon solvents.  This result was interpreted to mean that excitation energy is *transported* at a faster rate ($k_{SENS} > k_{DIF}$) in neat acetone as a result of *energy hopping* from triplet acetone to ground-state acetone. This conclusion was challenged (13) and reinterpreted as an effect on acetone *lifetime* due to quenching by hydrogen abstraction.  Although this objection is probably valid, the

---

$^†$  Unreactive, in this context, means that the solvent does not contribute any major pathway for deactivation of acetone triplet.

quantitative aspects of the work put forth in support of
reinterpretation of the energy hopping conclusion have, in
turn, been called to question (11, 14).

More recently (15), the evaluated rate constant for
singlet energy transfer from acetone to various aromatic
solutes in neat acetone were found to be about an order of
magnitude greater ($k \sim 10^{11}$ $M^{-1}$ $sec^{-1}$) than $k_{DIF}$. These
data were interpreted in terms of resonance interactions
between acetone molecules. This particular conclusion, however, would appear to be untenable in view of a more direct
measurement of a very low rate constant found for energy
transfer from singlet acetone to ground-state acetone (16)
($k_{HOP} \sim 5 \times 10^6$ $M^{-1}$ $sec^{-1}$).

## THE ACETONE EXCIMER: SOME EXPERIMENTAL OBSERVATIONS OF
## ACETONE PHOTOCHEMISTRY

In a general way, the net photochemistry of acetone in
unreactive solvents is very inefficient (17). In fact,
acetone is very unphotoreactive as the pure liquid, which
means that acetone, with respect to photoreactivity, qualifies
as an unreactive solvent for itself. These results contrast
somewhat with the observation of self-quenching processes in
other ketones (18). The only dimeric products isolated (in
less than 1% yield!) from photolysis suggest that excited
acetone may eventually attack the low concentration of acetone
enol. An important experiment relevant to the problem of the
acetone excimer was reported by Yang (19). It was observed
that acetone-$^{18}$O and acetone-$d_6$ undergo photochemical iso-
topic exchange to yield acetone and acetone-$d_6$-$^{18}$O. From a
combination of quenching studies and the observation of a
catalytic effect of water on the exchange reaction, it was

concluded that singlet acetone undergoes exchange *via* a
dimer intermediate, possibly 1 (equation [2]) and that triplet acetone may exchange *via* a photohydrate such as 2
(equation [3]).

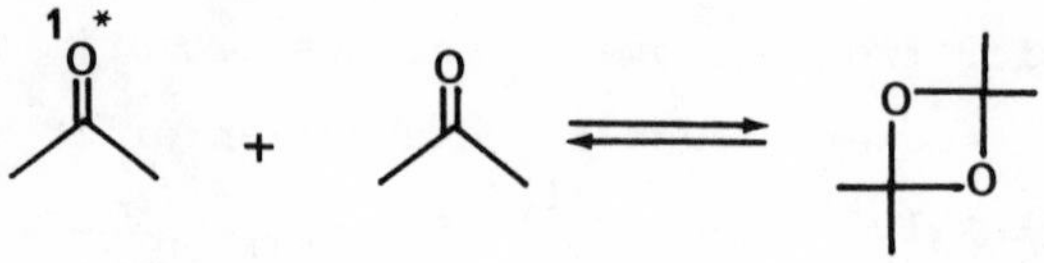

[2]

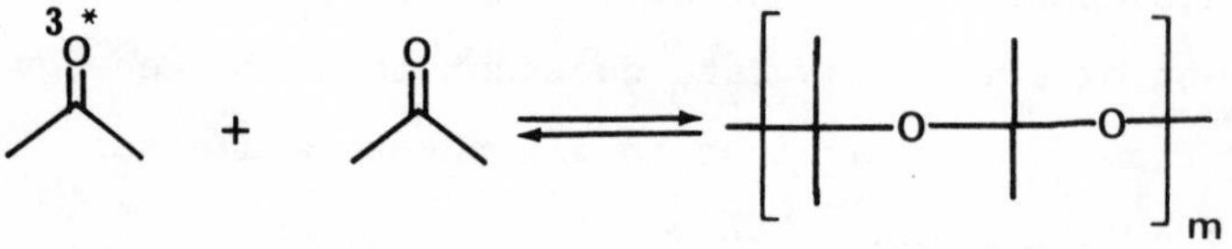

1

[3]

2

Since the lifetime of singlet acetone does not vary more
than a few percent (*i.e.*, it is constant within the reported
experimental error) upon proceeding from dilute solution to
pure liquid acetone, we must conclude that, if equation [2]
is correct (*vide supra*), 1 must be in reversible equilibrium
with $^1A^*$ and A and that 1 does not have any significant decay
pathways which involve quenching of electronic excitation.

THE ACETONE EXCIMER: A SUMMARY

To date, the published literature on the photophysics of
acetone is conflicting but recent work appears to be clarifying
some important points. First, both acetone fluorescence and
phosphorescence are observable in photochemically unreactive
solvents such as acetonitrile, acetone and Freon (11b). Next,
the measured lifetimes of both acetone fluorescence (15) and
phosphorescence (11) show no signs of self-quenching.
Finally, the only evidence for efficient interactions between
excited acetone and ground-state acetone is the photochemical
exchange reaction of acetone (equation [2]) (19). The latter

result requires a comparable energy for the proposed intermediate $\underline{1}$ and ($^1$A* + A) in order to be consistent with both
the sizable (~ 25%) (19) limiting quantum yield for exchange
and the lack of significant self-quenching of acetone
luminescence. This conclusion brings along with it the
requirement that the energy content of $\underline{1}$ relative to two acetone molecules is much higher than expected from bond-energy
considerations or that $\underline{1}$ is not a thermally-equilibrated species. Neither one of these requirements is consistent with
expectations, which suggest that $\underline{1}$ is an unusual species or
that another, undisclosed explanation of the exchange reaction
exists.

In effect, then, the experimental evidence for excimers
of acetone is weak at best.

TETRAMETHYL-1,2-DIOXETANE: IS IT A "CRYPTO" EXCIMER?

The energy diagram for acetone is shown in figure 2, and

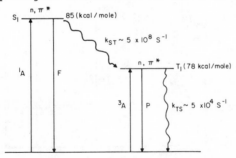

*Fig. 2  Energy diagram for acetone in a typical
unreactive solvent.*

the expected structures of $^1$A* and $^3$A* are shown in figure 3.
By use of the method of Kopecky (20), we have synthesized
tetramethyl-1,2-dioxetane, $\underline{3}$, an acetone "dimer" whose properties are interesting to compare to those of A*. Figure 4
shows the energy levels of $\underline{3}$, its transition state for

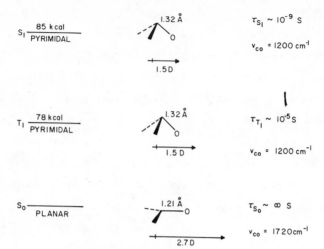

Fig. 3   *Schematic description of the excited-state*
*structure of acetone.   (Derived from values*
*for formaldehyde).*

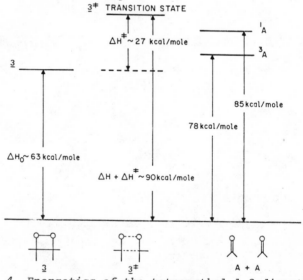

Fig. 4   *Energetics of the tetramethyl-1,2-dioxetane*
*and acetone systems.   The energy of two*
*ground-state acetones is considered as the*
*zero point.*

decomposition and the energies of $^1A*$ and $^3A*$; all energies are relative to two ground-state acetone molecules.

The points of interest are that the energy content of 3 (~65 kcal/mole) is comparable to that of (A* + A) and that, as 3 approaches the transition state for its decomposition, its energy actually rises above that of both $^1A*$ and $^3A*$. The nuclear structure of each "half" of 3 (considering the molecule as two acetone moieties) looks more reminiscent of A* than A. Indeed a simple MO bonding scheme (21) of 3 suggests that the O-O bond possesses two electrons in a $\pi$" orbital! These considerations lead to the notion that 3 possesses many of the attributes of an excimer of acetone: (i) high electronic energy content; (ii) nuclear structure comparable to A* and (iii) possession of $\pi*$ electrons.

Indeed, upon heating, 3 undergoes (22) efficient *intersystem crossing* to $^3A*$ + A as the major decomposition pathway! Thus, 3 exhibits one of the common and important properties of excimers; namely, the ability to fragment into two particles which are chemically identical except for electronic excitation (equation [1]). Surely, it is rather unorthodox to talk about a molecule which can be put into a bottle (3 is a kinetically-stable yellow solid at 25° C) as if it were an excimer. Yet 3 is a rather unorthodox molecule. Recall that in our own definition of an excimer, the key ideas were (i) electronically-excited molecular complex of definite stoichiometry; (ii) formed from two molecules of the same species and (iii) dissociated in the ground state. The only idea that is difficult to accept is the notion that 3 is an electronically-excited state of two acetone molecules. Conceptually, I believe the difficulty lies in the expectation that an electronically-excited state will have a lower state to which it can decay and thus emit radiation. Figure 5 indicates schematic energy diagrams for the "classic" excimer

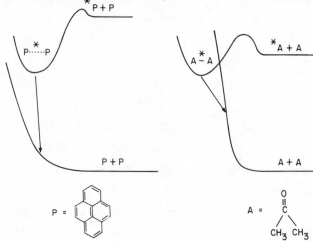

**Fig. 5** *A comparison of the pyrene excimer and the "crypto" excimer, in terms of simple energy diagrams.*

system, pyrene, and for the "crypto" excimer system, tetra-methyl-1,2-dioxetane. We consider here just the notion of a lowest excited-state surface. The obvious difference between the two systems is the position of the ground-state potential energy curve relative to the excited-state curve. Apparently for the case of **3** the curves either "cross" or, if they "do not cross", the "crypto" excimer A—*—A "jumps" from its original surface to the ground state. If we consent to this picture for describing the ground state and lowest excited state of **3**, we see that the relation of A—*—A to A* + A is very similar to that for P-*--P to P* + P. The major difference would appear to be the lack of any measurable radiative strength of A—*—A. This incredibly low rate of emission would have to be ascribed to a "world's record" Franck-Condon forbiddenness for the A—*—A ⟶ A + A + hν process. The latter situation seems a bit far-fetched, but it would be of interest to quantify the situation by making a calculation of the emission probability of the

A—$\overset{*}{\quad}$—A ⟶ A + A process. Indeed, the twisted triplet of ethylene (23) is a known (theoretical) case for which Franck-Condon forbiddenness is expected to imbue the excited molecule with an exceptionally long lifetime. In essence, the inability of tetramethyl-1,2-dioxetane to radiate makes the molecule a "foiled" electronically excited state, *i.e.*, instead of being metastable in the usual sense, it is kinetically stable at room temperature for all intents and purposes.

TETRAMETHYL-1,2-DIOXETANE AS A "MASKED" ELECTRONICALLY EXCITED

MOLECULE

The concept of a "masked" functional group has had a great impact on the art of organic syntheses. This notion identifies a chemical   functionality (the "masked" function group) as synthetically equivalent to another "target" functionality when the "masked" group is easily convertible to the target functionality under controllable conditions. In a true operational sense, we can extend this idea to consider tetramethyl-1,2-dioxetane as a "masked" electronically-excited state of acetone (24).

For example, we (22) have been able to trigger the fragmentation of A—$\overset{*}{\quad}$—A into A* + A by (i) mild heating, (ii) direct photoexcitation, (iii) singlet photosensitization and (iv) triplet photosensitization. In principle, thermal catalysis of the conversion of A—$\overset{*}{\quad}$—A into A* + A should be possible, but has not been achieved to date. The concept of A—$\overset{*}{\quad}$—A as a "masked" excited state of A* can be a powerful device for formulating novel experiments.

$^1$A* and $^3$A* are (by virtue of their relatively higher electronic excitation energies) choice candidates as electronic energy donors in energy-transfer experiments. However, acetone's weak absorption ($\varepsilon_{max} \sim 20$) and its high energy ($\lambda_{max} \sim 280$ nm) make energy transfer experiments involving a strongly absorbing acceptor technically impossible. Mild heating ($\sim 40 - 70^{O}$ C) of solutions of A—$*$—A produces A* (mainly triplet) in good yield. Thus heating of A—$*$—A in the presence of a strongly absorbing energy acceptor (Q) allows the straightforward study of the energy-transfer process, A* + Q $\longrightarrow$ A + Q*. Employing this technique, we [25] have been able to uncover an interesting long-range singlet-singlet energy transfer from $^{1}$A* to anthracenes and an unexpectedly fast triplet-singlet energy transfer from $^{3}$A* to heavy atom containing acceptors.

Of particular interest with respect to the acetone excimer problem, however, is a set of experiments which demonstrate electronic energy transfer from excited acetone to ground-state acetone.

## DIRECT CHEMICAL DEMONSTRATION OF ENERGY HOPPING IN ACETONE SOLVENT. EXCITATION DIFFUSION VERSUS EXCITATION MIGRATION

Imagine exciting a solution of acetone with light. The electronically-excited acetone molecules, A*, produced by absorption of the light, find themselves immersed in a sea of ground-state acetone molecules. Now suppose a quencher molecule exists at some distance, R, from the site of an A* molecule. The excitation of A* may be quenched by one of two pathways: (a) material or molecular diffusion of A* through solvent A molecules until an encounter with Q is achieved or (b) excitation diffusion by excitation hopping from A* to an adjacent A molecule, and continuation of the excitation hopping until an A* and Q are encounter partners. Schematically (figure 6), the competition between paths a and b are a

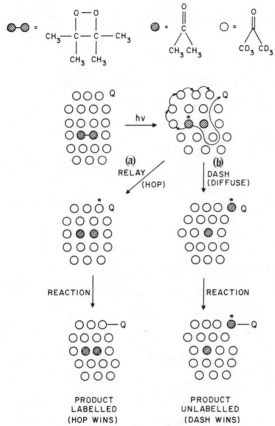

Fig. 6  Possibilities of energy transfer in liquid
acetone. An originally excited acetone molecule
(unlabelled, shaded circles) is generated in labelled
acetone solvent (open circles). The excited
unlabelled acetone can diffuse to the site of Q and
react with it (to produce an unlabelled acetone-Q
adduct). The excited, unlabelled acetone can also
transfer its excitation to a labelled acetone molecule in the solvent. The transfer act is irreversible
because of the much higher concentration of unlabelled
acetone. The excited, unlabelled acetone may then
react with Q to produce a labelled acetone-Q adduct.
Thus, the occurrence of two fundamentally different
mechanisms of a relay (hop) of energy through the
solvent or material diffusion (dash) of energy through
the solvent may be evaluated by analyzing for the
occurrence or absence of label in the acetone-Q adduct.

competition between a "random molecular walk" of A* to the
site of Q and an "electronic excitation relay" with solvent
A molecules serving as the relay team.

These ideas are part of a more general set of "null" or
degenerate photochemical processes, which have some special
theoretical and practical implications (equation [4]). A
convincing demonstration of

$$A* + A \rightleftharpoons A + A* \qquad [4]$$

the occurrence of reaction [4] is a particularly intriguing
and obvious experimental challenge. As a first step in
designing experimental strategy to meet this challenge, consider the notion of *labelled* acetone (A-$d_6$) solvent that contains a number of isotropically distributed, but electronically-
excited unlabelled acetone molecules (A*-$h_6$). Now consider
as a specific quencher an ethylene capable of forming an oxetane with A*. *If* we could achieve the initial conditions
described above (A*-$h_6$ in A-$d_6$ at t = 0) then we can test for
excitation diffusion by *simply analyzing for oxetane containing the elements of A-$d_6$*. Figure 7 outlines the key
processes where $k_{HOP}$ is the rate constant for energy hopping
*via* excitation hopping, and $k_{OX}^H$ and $k_{OX}^D$ are the rate constants
for formation of oxetanes for A*-$h_6$ and A*-$d_6$, respectively.

Now the problem boils down to the specific generation of
A*-$h_6$ in solvent A-$d_6$. This is, as the reader may have
guessed by now, an easy task if one employs 3 as a source of
A*-$h_6$. Selective chemiexcitation can be used to generate
uniquely A*-$h_6$ from 1 in A-$d_6$ solvent! Furthermore, since we
know from earlier studies (26) that 1,2-dicyanoethylene (DCE)
reacts efficiently only with $^1$A and that 1,2-diethoxyethylene
(DEE) reacts efficiently only with $^3$A, we can measure *both*
singlet excitation hopping and triplet excitation hopping.

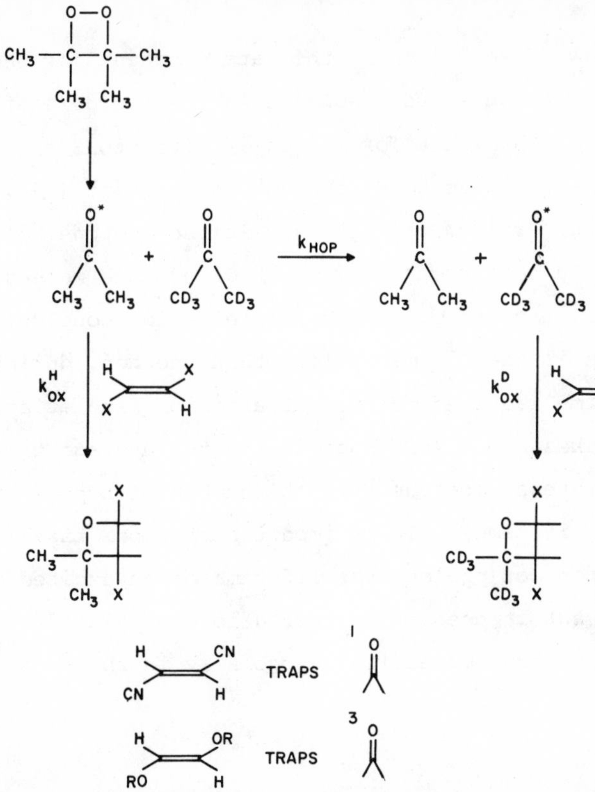

Fig. 7 Flow diagram of the titration of unlabelled
and labelled electronically-excited acetone molecules.
1,2-Dicyanoethylene reacts specifically with singlet
acetone and 1,2-diethoxyethylene reacts specifically
with triplet acetone. The occurrence or nonoccurrence
of deuterium in the oxetane product allows calculation
of the extent of energy hopping from initially excited
unlabelled acetone (produced from 1) to the labelled
acetone solvent.

Experimentally, it is found that at 0.1 $M$ DCE, 5% of the
oxetane from $^{1}A$ contains $d_6$ and at 0.1 $M$ DEE, 10% of the
oxetane from $^{3}A$ contains $d_6$. Since $k_{OX}^{H}$, which must be
essentially identical to $k_{OX}^{D}$, is known for both $^{1}A$ + DCE and
for $^{3}A$ + DEE, we can easily calculate $k_{HOP}^{S}$ and $k_{HOP}^{T}$ for
singlet hopping and for triplet hopping, respectively.

The results of the evaluation of $k_{HOP}^{S}$ and $k_{HOP}^{T}$ are that,
within the experimental error, both are equal to

$\simeq 3 \times 10^6 \ M^{-1} \ \text{sec}^{-1}$. Thus, the rate constant for energy diffusion through acetone solvent is 3 - 4 orders of magnitude smaller than that for molecular diffusion ($k_{DIF} \simeq 10^{10} \ M^{-1} \ \text{sec}^{-1}$ in acetone) through the solvent (27). In terms of the "random walker", an excited acetone molecule would easily beat the "relay team" of acetone molecules in the microscopic race to Q! Another interesting conclusion from these data is that $^1A$ makes less than one hop, during the average time required for deactivation of singlet acetone and that $^3A$ makes about *1000* hops during the average time required to deactivate an acetone triplet (number of hops = rate of hopping × lifetime). It is important to note that energy hopping does not prolong the lifetime of an excited acetone molecule, but it does allow excitation to "visit" a larger number of acetone molecules, especially in the triplet.

CONCLUSION

In this paper, we have attempted to put forth some novel viewpoints concerning standard notions of excimers and electronic excitation. These ideas are put forth not in a spirit of heresy or proselytism, but with the benign, heuristic purpose of stimulating an unorthodox point of view which will suggest experiments and concepts which do not arise naturally from our standard ways of thinking about electronically-excited molecules. In fact, this methodology has provided us with a rather fruitful intellectual basis for designing experiments which probably would not have otherwise occurred to us. We therefore believe it to have achieved some measure of validity for us and we hope others may also find some utility in this way of thinking.

## ACKNOWLEDGEMENTS

This work was generously supported by the Air Force Office of Scientific Research (Grant AFOSR-74-2589) and the National Science Foundation (Grants NSF-GP-40330x and NSF-GP-74-26602x).  This work was executed by Dr. Peter Lechtken, Dr. Hans-Christian Steinmetzer, Dr. Gary Schuster and Dr. Ahmad Yekta.  In addition to the substantial intellectual contributions made by these workers, the author thanks Professor Richard Bersohn for his many helpful discussions and contributions to the ideas in this work.

REFERENCES

1. J.B. Birks, "Photophysics of Aromatic Molecules." Wiley-Interscience, New York, 1970.

2. Th. Förster, *Angew. Chem. Int. Ed.* *8*, 261 (1969).

3. T.F. Lin, S.D. Christian and H.F. Affsprung, *J. Phys. Chem.* *71*, 968 (1967).

4. a) K. Fukui, *Topics Curr. Chem.* *15*, 1 (1970); K. Fukui, *Accounts Chem. Res.* *4*, 57 (1971); (b) W. Herndon, personal communication.

5. A. Halpern and W.R. Ware, *J. Chem. Phys.* *54*, 2171 (1971).

6. R.F. Borkman and D.R. Kearns, *ibid.* *44*, 945 (1966); E.H. Gilmore, G.E. Gibson and D.S. McClure, *ibid.* *20* , 829 (1952); *23*, 399 (1955).

7. M. O'Sullivan and A.C. Testa, *J. Amer. Chem. Soc.* *90*, 6245 (1968).

8. G.D. Renkes and F.S. Wettack, *ibid.* *91* , 7514 (1969).

9. Extensive attempts in our laboratories at Columbia failed to produce any evidence for the emission reported in ref. 7.

10. A.C. Testa, personal communication; M. O'Sullivan and A.C. Testa, *J. Amer. Chem. Soc.* *92*, 5842 (1970).

11. a) G. Porter, R.W. Yip, J.M. Dunston, A.J. Cessna and S.E. Sugamori, *Trans. Faraday Soc.* *67*, 3149 (1971); b) N.J. Turro, H.C. Steinmetzer and A. Yekta, *J. Amer. Chem. Soc.* *95*, 6468 (1973).

12. R.F. Borkman and D.R. Kearns, *ibid.* *88* , 3467 (1966).

13. P.J. Wagner, *ibid.* *88*, 5672 (1966).

14. The confusion results from a) indirect methods of evaluating triplet lifetimes and b) differing reactivity

toward quenching of acetone triplet in different solvents.

15. A.J. Robinson, M.A.J. Rodgers, J.P. Keene and C.W. Gilbert, *J. Photochem.* $\underline{1}$, 379 (1972/73).

16. P. Lechtken and N.J. Turro, *Angew. Chem. Int. Ed.* $\underline{12}$, 314 (1973).

17. a) E.J. Bowen and E.L.A.E. de la Praudiere, *J. Chem. Soc.* 1503 (1934); b) P.E. Frankenburg and W.A. Noyes, *J. Amer. Chem. Soc.* $\underline{75}$, 2847 (1953); c) R. Picek and E.W.R. Steacie, *Can. J. Chem.* $\underline{33}$, 1304 (1955); d) K. Pfordt and G. Leuschner, *Ann. Chem.* $\underline{646}$, 23 (1961); e) J.T. Przybytek, S.P. Singh and J. Kagan, *Chem. Commun.* 1224 (1969); f) Review: T. Berces *in* "Comprehensive Chemical Kinetics." Vol. 5, (C.H. Bamford and C.F.H. Tripper, eds.) Elsevier, London, 1972.

18. a) O.L. Chapman and G. Wampfler, *J. Amer. Chem. Soc.* $\underline{91}$, 5390 (1969); b) A. Yekta and N.J. Turro, *Mol. Photochem.* $\underline{3}$, 307 (1972); c) L. Giering, M. Berger and C. Steel, *J. Amer. Chem. Soc.* $\underline{96}$, 953 (1974).

19. N.C. Yang, W. Eisenhardt and J. Libman, *ibid.* $\underline{94}$, 4030 (1972).

20. K.R. Kopecky and C. Mumford, *Can. J. Chem.* $\underline{47}$, 709 (1969).

21. D.R. Kearns, *Chem. Rev.* $\underline{71}$, 395 (1971).

22. N.J. Turro, P. Lechtken, N.E. Schore, G. Schuster, H.C. Steinmetzer and A. Yekta, *Accounts Chem. Res.* $\underline{7}$, 97 (1974).

23. A.J. Merer and R.S. Mulliken, *Chem. Rev.* $\underline{69}$, 639 (1969).

24. N.J. Turro and P. Lechtken, *Tetrahedron Letters* 565 (1973).

25. N.J. Turro, P. Lechtken, G. Schuster, J. Orell, H.C. Steinmetzer and W. Adam, *J. Amer. Chem. Soc.* $\underline{96}$, 1627 (1974).

26. N.J. Turro, *Pure Appl. Chem.* $\underline{27}$, 697 (1972).

27. It is interesting to note that for relatively rigid
aromatic hydrocarbons, excitation migration is
thought to be _faster_ than diffusion controlled:
for example, see J.W. Eastman, E.J. Smutny and
G.M. Coppinger, *J. Chem. Phys.* *53*, 4346 (1970);
J.W. van Loben Sels and S.T. Dubois, *Molec. Crystals*
*4*, 33 (1968).

# COMPLEXES OF DIPOLAR EXCITED STATES

# AND SMALL POLAR MOLECULES

EDWIN A. CHANDROSS

*Bell Laboratories*

*Murray Hill, New Jersey 07974*

## INTRODUCTION

The formation and properties of exciplexes involving
exciton interactions (excimers) and/or charge-transfer interactions is a well characterized phenomenon, pioneered by the
work of Weller's and Mataga's laboratories. The work described in this paper involves exciplexes which are stabilized
by more conventional solvent-solute interactions, in contrast
to the usual charge-transfer stabilization of heterocomplexes.
It seems clear that stoichiometric multicomponent complexes
containing one dipolar excited molecule and one or more small
polar molecules are formed. In one case, interaction with
one or perhaps two molecules of dimethylformamide is enough
to change the nature of the lowest excited state from a
localized $\pi,\pi^*$ state to a charge-transfer state, a phenomenon
previously realized only in pure polar solvents.

Our studies began with the intramolecular exciplex of
N,N-dimethyl-3-(1-napthyl)propylamine and much of this has

been published previously (1). We found it advantageous to study simpler systems where the changes in geometry between ground and excited states were as small as possible. The most interesting species studied thus far is 9-(4-dimethyl-aminophenyl)anthracene. The results obtained are frequently reminiscent of those reported by Lumry, et al. (2) for the interaction of singlet excited indole with, i.a., dioxane or acetonitrile for which stoichiometric complex formation was suggested. We have avoided any system which offers the possibility of hydrogen bonding in either state as we feel that it adds the possibility of an unknown quantity to the interactions.

The results obtained with the intramolecular exciplex were explained on the basis of stoichiometric complex forma-tion between the exciplex, which has a large dipole moment (ca. 12 D), and small dipolar molecules, analogous to the dimers formed by species such as acetonitrile. We suggested that the primary interaction was between the positive end of the exciplex dipole with the negative end of the small polar molecule. The experiments on other species are in accord with these conclusions.

In spite of the fact that the association of dipolar molecules is expected on the basis of simple classical con-cepts, there has been very little work done on molecules in solution. The subject has been discussed by Treiner, Skinner and Fuoss (3) who concluded that dipole-dipole association occurs for p-nitroaniline, m-nitrophenol and pyridinium dicyanomethylide and that the energies of association appear to depend on the square of the dipole moment, as expected. The heat of association is about 3-5 kcal mole$^{-1}$ if $\Delta S$ is assumed to be in the range of 10 - 15 eu; the equilibrium constants were found to be close to 1. Similar numbers have been suggested for the association of nitriles (4 - 7).

The specific solvation of tetralkylammonium ions by dipolar molecules such as phosphine oxides has been shown to occur by Gilkerson and Ezell (8). This is very similar to the interaction we postulate to explain our results.

The conclusions drawn in our first paper were critized by Selinger (9) who denied the existence of the type of complexes we suggested. The work presented here points clearly to the validity of the original hypotheses.

It is well known that the fluorescence spectrum of a species whose molecules have much larger dipole moments in the excited state is dependent on solvent polarity and polarizability. The relationship is described by the theories of Lippert (10) and Mataga (11) which, however, assume no specific interaction. We seek to demonstrate that many small dipolar molecules exhibit specific molecular interactions, oftentimes at concentrations which are low enough not to have any large effect on solvent polarity. The important feature in these studies is not the change in intensity but rather the fluorescence band shape, frequently indicative of more than one component. It is, of course, preferable to have a series of spectra which has an isoemissive point and is thus clearly indicative of a two-component system. The fluorescence spectra of the intramolecular exciplex published previously were chosen to emphasize this. The chances for observing simple interactions are best at the lowest concentrations of added dipolar species which avoids any significant change in the bulk solvent properties. This implies a preference in these studies for strongly interacting dipolar species and dipolar excited states of relatively long lifetimes.

The nature of the solvent used for these studies is quite important. In order to approach gas-phase conditions we used hydrocarbon solvents with low dielectric constants (e.g., methylcyclohexane, $\varepsilon = 2.0$, n = 1.42). Selinger has pointed out the desirability of working at constant dielectric strength and suggested the use of ethyl acetate. However, one must be careful to note the "leveling effect" on complex formation of a somewhat polar, moderately Lewis-basic solvent which interacts less strongly with the polar excited state but nevertheless overwhelms the added polar species by virtue of its much higher concentration. This is quite analogous to the dissociation of protic acids in water where many appear to be equally strong but are found to be easily differentiable in strength when dissolved in a much less basic solvent.

As mentioned, much of the work on the intramolecular naphthalene-amine exciplex has been published. A few points deserve further discussion. We consider first the case of quenching by triethylamine. The fluorescence spectra of the exciplex at various concentrations of the amine are shown in figure 1; figure 2 gives the Stern-Volmer plot which yields $k_Q = 9.0 \times 10^7$ l mole$^{-1}$ sec$^{-1}$. The dielectric constant of triethylamine is 2.4; its refractive index is 1.40. It is reasonable to assume that these properties of the solvent will change in a linear manner upon addition of amine and, at the concentrations used, the change is small indeed. Further, the addition of toluene ($\varepsilon = 2.4$, n = 1.50) has virtually no effect on the fluorescence of the exciplex and the effect of 1,2-dimethoxyethane ($\varepsilon = 6$) is far smaller than that of triethylamine. Contrary to the assertions of Selinger, it is clearly beyond question that the interaction between tri-ethylamine and the exciplex is a specific one and is not simply related to a change in bulk solvent dielectric

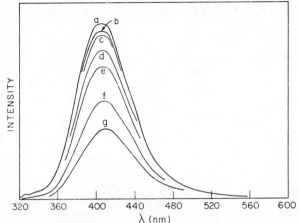

Fig. 1 *Quenching of the exciplex fluorescence in*
*methylcyclohexane solution by triethylamine.*
*1% = 0.073 M = 0.93 mole percent. a, none;*
*b, 0.2%; c, 0.5%; d, 1%; e, 2%; f, 5%;*
*g, 9% triethylamine.*

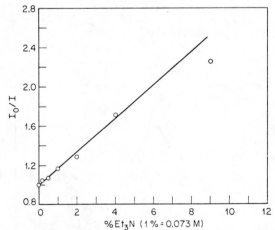

Fig. 2 *A Stern-Volmer plot for triethylamine*
*quenching.*

properties. The fact that the fluorescence is simply quenched
rather than shifted by triethylamine is undoubtedly a
consequence of its low ionization potential.

A further demonstration of different degrees of specific
interaction rather than bulk solvent properties is apparent

191

in figures 3 and 4 which show the effect of propionitrile

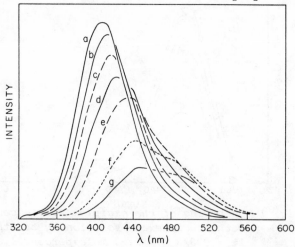

Fig. 3 The effect of propionitrile on the exciplex
fluorescence in methylcyclohexane solution.
a, none; b, 0.5%; c, 1%; d, 2%; e, 4%;
f, 9%. 1% = 0.14 M = 1.8 mole percent
propionitrile.

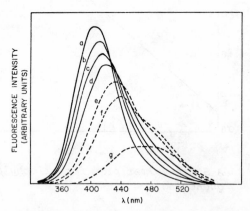

Fig. 4 The effect of hexamethylphosphoramide on the
exciplex fluorescence in methylcyclohexane
solution. a, none; b, 0.1%; c, 0.2%;
d, 0.4%; e, 1%; f, 2%; g, 9%. 1% = 0.63M
= 0.75 mole percent hexamethylphosphoramide.

($\varepsilon$ = 27, n = 1.36) and hexamethylphosphoramide (HMPA, $\varepsilon$ = 30, n = 1.46), two liquids of comparable polarity and dielectric constant. It is clear that HMPA, an excellent solvating agent for cationic centers, is far more effective than the nitrile. The change in the fluorescence spectrum may be treated as a quenching process. A Stern-Volmer treatment of these spectra (380 nm) gives $k_Q$ = 4.3 × $10^8$ 1 mole$^{-1}$ sec$^{-1}$ for propionitrile, and 1.0 × $10^{10}$ 1 mole$^{-1}$ sec$^{-1}$ for HMPA. The latter plot curves upward for concentrations of HMPA above 0.06 $M$, suggesting interaction with more than one molecule. HMPA itself probably associates at moderate concentrations. As an aside, it may be noted that acetonitrile is only modestly soluble in aliphatic hydrocarbons; this is undoubtedly a consequence of dipolar association.

Finally, we return to the fluorescence band shape. As noted earlier, it would be preferable to generate a series of spectra having an isoemissive point in order to establish the existence of a two-component system. Unfortunately in these experiments it is usually not possible to do this, probably because there are complexes involving two or three small dipolar molecules which form readily and compete with the 1:1 complex. However, by examining the band shape it is frequently easy to infer the existence of a two or more component system, complicated further by a small bulk-solvent-induced shift, which differs from the simple red shift expected from the bulk solvent-solute interaction as treated in the theories of Lippert and Mataga. Figure 5 shows a series of fluorescence spectra generated by the addition of dimethylformamide ($\varepsilon$ = 37, n = 1.43) to the exciplex in benzene solution. The intensities are not corrected for dilution (a simple linear correction in these experiments) and there is no isoemissive point. The changing band shape

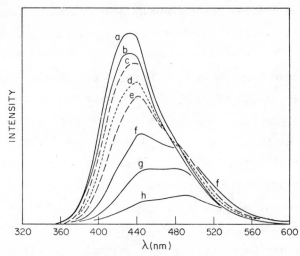

Fig. 5  *The effect of N,N-dimethylformamide on the*
*exciplex fluorescence in benzene solution.*
*a, none; b, 0.2%; c, 1%; d, 2%; e, 4%;*
*f, 9%; g, 17%; h, 28%. 1% = 0.13 M = 1.6*
*mole percent dimethylformamide.*

is indicative of the point, however.  Curve f appears to
have a component whose maximum is slightly red-shifted from
the curves at lower concentration while most of the intensity
lies in the species having a maximum at 480-500 nm.  Curve g
with its flat top is also suggestive of two (or three) components.  A Stern-Volmer plot (410 nm) of these spectra gives
a "quenching rate constant" of $9.3 \times 10^8$ 1 mole$^{-1}$ sec$^{-1}$,
about 10% that of a diffusion-controlled process.

Similar experiments were carried out with acetonitrile
in benzene solution.  Figure 6 shows the fluorescence spectrum
in a 4:1 mixture of acetonitrile:benzene along with the
difference spectrum obtained from the fluorescence measured
in 1M acetonitrile and that in pure benzene.  Even though
one may not accept such spectra as quantitative evidence, the
similarity between the two is striking and suggests that the
excited molecule is completely solvated in the 1 M solution.

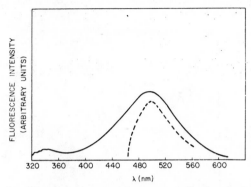

*Fig. 6    Fluorescence spectrum of the intramolecular exciplex in 80% acetonitrile - 20% benzene (full line) and the difference spectrum (dashed line, X5) obtained from spectra in pure benzene and a 1 M solution of acetonitrile in benzene.*

We next attempted to study the solvent dependence of 4-dimethylaminobenzonitrile (DMABN) in a series of similar experiments.  The behavior of this species has been studied more recently by Mataga (12) who has reviewed the area and suggested the formation of specific polar interactions as we did.  The kinetic behavior of this system has also been examined by Struve and Rentzepis (13) who worked with individual pure solvents.

Lippert (14) suggested that the marked effect of solvent polarity upon the fluorescence of DMABN could be explained by postulating that the localized benzenoid ($^1L_b$) state is the emitter in nonpolar solvents and that there is a charge-transfer (quinoid) state at somewhat higher energy.  In polar media, the latter state is stabilized and becomes the lower of the two.  A recent paper has dealt with this change in terms of molecular geometry (15).

The spectrum of DMABN as a function of added

propionitrile (methylcyclohexane solvent) is shown in
figures 7 and 8. At low concentrations of nitrile, there is

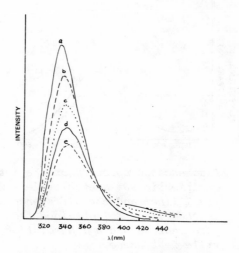

Fig. 7   *Fluorescence spectrum of DMABN in methylcyclohexane solution (7 × 10⁻⁵ M, excited at
298 nm) as a function of added propionitrile
(a, none;  b, 1% (0.14 M);  c, 2%;  d, 3%;
e, 4%.*

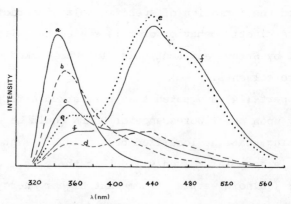

Fig. 8   *Fluorescence spectrum of DMABN in methylcyclohexane solution with added propionitrile
(a, none;  b, 2%;  c, 4%;  d, 9%;  e, 17%;
f, 29%;  the intensity for curves e and f has
been multiplied by five.*

primarily a diminution in intensity along with a very small red shift of the maximum and an increase in emission at the red end of the spectrum. However, high concentrations of propionitrile (figure 8) lead first to a spectrum with two components and then to a spectrum which has three components. The spectrum of uncomplexed DMABN still seems to persist, with only a moderate red shift, as the new bands gain in prominence. The fluorescence intensity decreases steadily as the polarity increases. If this system is treated as a Stern-Volmer quenching (in the low-concentration region), we found that $k_Q = 3 \times 10^9$, or about 10% of the diffusion controlled rate. The lifetime was estimated as 3 nsec by Berlman's method. Dimethylformamide showed even stronger interactions; we found $k_Q = 1 \times 10^{10}$.

There is thus no question that singlet excited 4-dimethylaminobenzonitrile interacts strongly with small polar molecules on a 1:1 basis. Mataga's studies ruled out ground-state complexes with nitriles and we found that moderate concentrations of DMF had no significant effect on the absorption spectrum. It is difficult to make more detailed studies because the lifetime is so short that moderate concentrations of polar species, which do change the bulk dielectric constant appreciably, are required.

We next sought a species similar to the aminobenzonitrile which might show simpler behavior and we chose 9-(4-dimethylaminophenyl)anthracene where the cyano group of the small molecule has been replaced by an anthracene nucleus which serves as an "electron acceptor". Sometime after this work was begun, a related study was published by Mataga's (16) group.

There are, *a priori*, a few comments that we can make about the geometry of this anilinoanthracene, based on

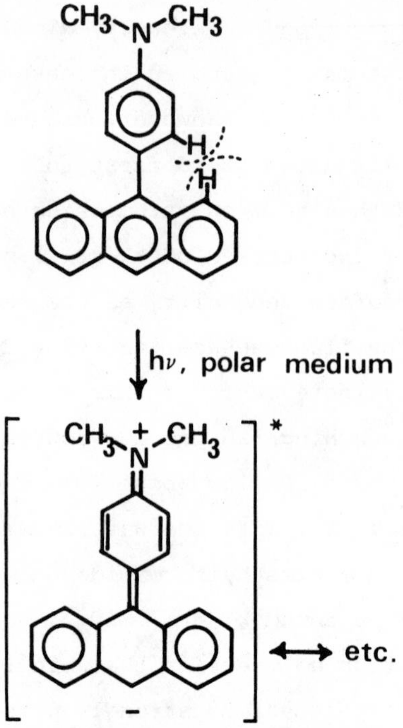

$h\nu$, polar medium

$\longleftrightarrow$ etc.

analogy to 9-phenyl- and 9,10-diphenylanthracene. Because
of steric interference between the *peri* hydrogens and those
at the *meta* positions of the phenyl moiety, there is a large
($\sim 65°$) twist between the planes of the two rings in the
ground state. The interaction in the excited state must be
greater as the mono- and di-phenylated anthracenes show some
distortion in the usual mirror-image symmetry and a somewhat
greater than usual Stokes shift between the two 0-0 bands.
Mataga's studies established that the lowest lying excited
state of the anilinoanthracene is the localized anthracene
state but that in polar solvents a nearby charge-transfer
state becomes lower in energy and large Stokes shifts are
observed. The situation is very reminiscent of 9,9'-bianthryl
as described by Lippert (17). A theoretical analysis of
this situation where a locally-excited or a charge-transfer
state can be lower in energy, depending on the environment,

198

has been given by Beens and Weller (18).

Mataga's studies, using the solvent-induced shift of the fluorescence, indicated that the change in dipole moment between $S_0$ and $S_1$ is 18 D in polar solvents. We have found that the fluorescence of this anilinoanthracene shows striking effects upon the addition of small quantities of polar solvents which can be explained only by specific interactions rather than a bulk-solvent-induced shift.

The fluorescence of the anilinoanthracene in nonpolar solvents looks very much like that of 9-phenylanthracene and is attributed to excitation localized on the anthracene nucleus. This spectrum is shown in figure 9 which also shows

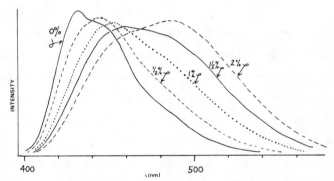

Fig. 9   *Fluorescence spectrum of the anilinoanthracene in methylcyclohexane solution (7 × 10⁻⁵ M,
excited at 313 nm) as a function of added
DMF (none, ½, 1, 1½ and 2%; 1% = 0.13 M).*

the changes induced by the addition of small quantities of N,N-dimethylformamide. The fluorescence was excited by 313 nm light because the absorption spectrum is featureless in this region and it is possible to avoid effects which could be caused by small shifts in band maxima in the more usual 365 nm excitation region. All solutions were deaerated by a stream of nitrogen. The polar species was added in successive portions and the solution was deaerated after each addition. The optical density at the excitation wavelength is about 0.05 cm$^{-1}$ and the fluorescence intensity is approximately a linear function of concentration in this region. Intensities were corrected for dilution by adjusting the fluorimeter gain.

The changes caused by the addition of DMF are of the two types mentioned earlier. The maximum is shifted to the red but, more importantly, the change in band shape indicates the presence of at least two emitting species.

We would obviously like to separate the changes in the fluorescence which are due to bulk dielectric effects from those which are due to specific interactions. This can be done to some extent by comparing the spectra involving strongly interacting species with those found for another solution of approximately the same dielectric constant. Thus, we have used o-dichlorobenzene ($\varepsilon$ = 9.9) to raise the dielectric constant of methylcyclohexane, because it shows much weaker specific interactions. We have ignored the refractive index portion of the Lippert-Mataga equation as it makes a much smaller contribution than the dielectric constant term.

The dielectric constants of the mixtures used were calculated by assuming the equation:

$$\varepsilon = \sum_i \text{Molarity}_i \; \frac{\varepsilon_i \text{ (pure)}}{\text{Molarity}_i \text{ (pure)}} = \sum_i \text{Volume fraction}_i \varepsilon_i$$

Figure 10 shows the spectrum of the anilinoanthracene in methylcyclohexane and the changes induced by the addition of 2% (0.26 $M$) DMF or 10% o-dichlorobenzene; the dielectric constants are 2.02, 2.66 and 2.61, respectively. The change caused by dichlorobenzene is simply a red shift of the band.

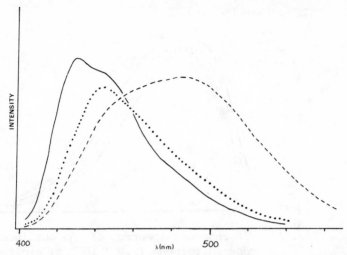

Fig. 10  Fluorescence spectrum of the anilinoanthracene in methylcyclohexane (as in figure 9)
and with 2% DMF (dashed line, ε = 2.66) or
10% o-dichlorobenzene (dotted line, ε = 2.61
added.

DMF has a much larger effect;  the band shape is altered
drastically in addition to the red shift.

Now, compare the effect caused by the addition of DMF
to a toluene solution of the anilinoanthracene (figure 11)
with the same experiment in methylcyclohexane solution
(figure 9).  The fluorescence spectrum is shifted to the red
in toluene but the change caused by DMF is now much less than
that observed in methylcyclohexane solution.  The isoemissive
point observed for the toluene solution looks nice but is
probably not real.  Further studies of this type should be
done at constant dielectric strength.  A more striking contrast is shown by figures 12 and 13.  In the first case, the
DMF concentration, using toluene and methylcyclohexane solvents, has been adjusted to give the same dielectric constant.
The band shapes are obviously different and the spectrum in
methylcyclohexane is shifted further to the red.  In figure 13

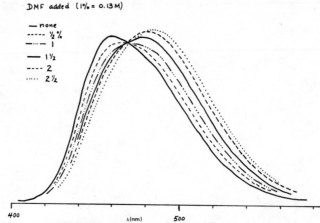

Fig. 11    Fluorescence spectrum of the anilinoanthracene in toluene ($1.0 \times 10^{-4}$ M, excited at
313 nm) with added DMF (none, ½%, 1%, 1½%,
2%, 2½%; 1% = 0.13 M).

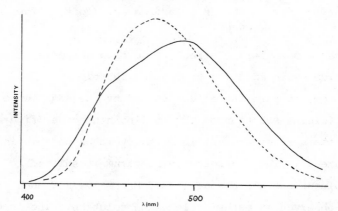

Fig. 12    Fluorescence spectrum of the anilinoanthracene in methylcyclohexane solution (full
line, as in figure 9) containing 2½%
DMF, ε = 2.84, and the spectrum in toluene
solution (dashed line, as in figure 11)
containing 1½% DMF, ε = 2.81.

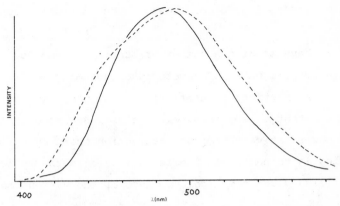

*Fig. 13   Fluorescence spectrum of the anilinoanthra-
cene in methylcyclohexane solution with
$2\frac{1}{2}$% DMF ($\varepsilon$ = 2.84, dashed line, conditions
as in figure 9) and in toluene with $2\frac{1}{2}$%
DMF (full line, $\varepsilon$ = 3.19, conditions as in
figure 11).*

we have the same amount of DMF present in both solvents but
the dielectric constant of the toluene solution (3.19) is
significantly greater than that of the methylcyclohexane
solution (2.84).   In spite of this difference, the spectrum
in the latter, less polar solution is shifted further to the
red.   All of these observations can be explained on the basis
of a solvent leveling effect discussed above.   The toluene
competes more effectively with DMF to solvate the excited
molecule than does methylcyclohexane.   The fact that the
spectrum is red-shifted further in the solution with the
lower dielectric constant is conclusive evidence for specific
interaction between the excited anilinoanthracene and DMF.

The changes in the fluorescence spectrum of the anilinoanthracene caused by DMF can also be treated as a Stern-Volmer
quenching.   A linear plot (at 420 nm) is obtained for 2% or
less DMF in toluene solution;   it curves upward at higher
concentrations.   The value of $k_Q\tau$ = 3.5.   We estimated $\tau$ by

the oxygen quenching technique and found it to be *ca.* 1 nsec; thus $k_Q$ is ~ $3.5 \times 10^9$ or about 1/3 the diffusion-controlled rate.

We have studied a number of other polar molecules in similar experiments. Hexamethylphosphoramide was found to exhibit even stronger interactions than DMF; the quenching rate constant is approximately the diffusion-controlled value.

The relative degree of interaction shown by various species in toluene solution is indicated in the following list which also gives the dipole moments of the various molecules.

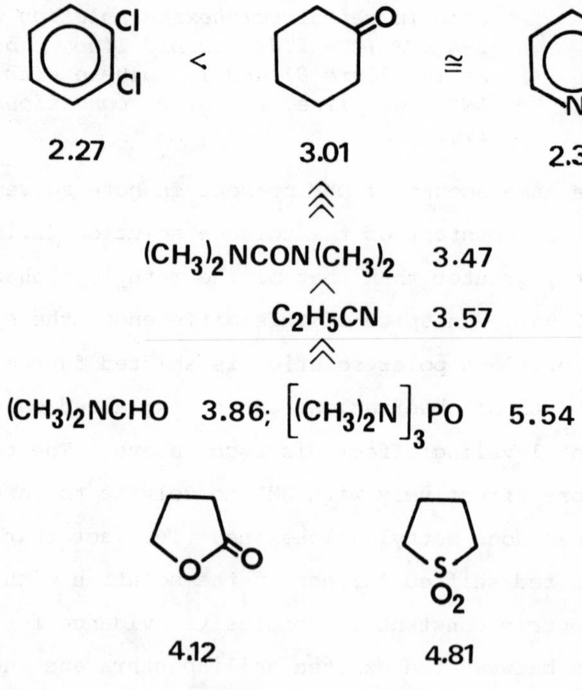

2.27          3.01          2.37

$(CH_3)_2 NCON(CH_3)_2$   3.47

$C_2H_5CN$   3.57

$(CH_3)_2 NCHO$   3.86;   $[(CH_3)_2N]_3 PO$   5.54

4.12          4.81

The conclusions drawn from these studies are essentially those drawn initially from the experiments with the intra-molecular naphthalene-amine exciplex. The specific inter-actions with small polar molecules seem to involve the positive end of the dipole of the excited molecule. This is quite reasonable; the positive ends of these dipoles are probably predominantly localized on the amino nitrogen while the negative ends are more diffuse. The species which give the strongest interaction are those which are best at stabilizing positive charges. There is not any significant possibility of charge-transfer interactions being important in the binding energy. The species studied are known to be both poor electron donors and acceptors. A possible exception to this is tetramethylene sulfone which might have some electron affinity. There does not seem to be any evidence for base strength being involved; pyridine interacts weakly and tetramethylurea only moderately.

It would be nice to have some quantitative data as to the binding energy in these complexes. We predict that they will fall in the area of 3-10 kcal mole$^{-1}$. The few studies of temperature dependence carried out thus far have been uninformative. Since the activation energy for diffusion is also about 3 kcal mole$^{-1}$, it is not clear that such studies will be useful. The time dependence of various parts of each fluorescence spectrum should also be interesting.

ACKNOWLEDGEMENTS

The experimental work on the intramolecular exciplex and DMABN was carried out by Dr. Harold T. Thomas, presently at Eastman Kodak. Both he and Dr. G.N. Taylor are thanked for stimulating dicussions of various aspects of this work.

REFERENCES

1. E.A. Chandross and H.T. Thomas, *Chem. Phys. Letters* *9*, 397 (1971).

2. M.S. Walker, T.W. Bednar and R. Lumry, *J. Chem. Phys.* *47*, 1020 (1967); also "Molecular Luminescence" (E. Lim, ed.) pp. 135-153, W.A. Benjamin, Inc., New York, 1969.

3. C. Treiner, J.F. Skinner and R.M. Fuoss, *J. Phys. Chem.* *68*, 3406 (1964).

4. W. Bannhauser and A.F. Fluckinger, *ibid.* *68*, 1814 (1964).

5. A.M. Saum, *J. Polymer Sci.* *42*, 57 (1960).

6. F.E. Murray and W.G. Schneider, *Can. J. Chem.* *33*, 797 (1955).

7. J.D. Lambert, G.A.H. Roberts, J.S. Rowlinson and V.J. Wilkinson, *Proc. Roy. Soc. A* *196*, 113 (1949).

8. W.R. Gilkerson and J.B. Ezell, *J. Amer. Chem. Soc.* *87*, 3812 (1965).

9. B.K. Selinger and R.J. McDonald, *Aust. J. Chem.* *25*, 897 (1972).

10. E. Lippert, *Angew. Chem.* *73*, 695 (1961).

11. See, for example, N. Mataga and T. Kubota, "Molecular Interactions and Electronic Spectra." Chap.6. Marcel Dekker, New York, 1970.

12. N. Nakashima and N. Mataga, *Bull. Chem. Soc. Japan* *46*, 3016 (1974).

13. W.S. Struve and P.M. Rentzepis, *J. Chem. Phys.* *60*, 1533 (1974).

14. E. Lippert, W. Luder and H. Boos, "Proc. 4th Internationa Meeting on Molecular Spectroscopy." (Bologna, 1959). p. 442. Macmillan, New York, 1962.

15.  K. Rotkiewicz, K.H. Grellmann, and Z.R. Grabowski,
     *Chem. Phys. Letters* *19*, 315 (1973).

16.  T. Okada, T. Fujita, M. Kubota, S. Masaki, N. Mataga,
     R. Ide, Y. Sakata and S. Misumi, *ibid.* *14*, 563 (1972).

17.  F. Schneider and E. Lippert, *Ber. Bunsenges. Phys. Chem.*
     *72*, 1155 (1968).

18.  H. Beens and A. Weller, *Chem. Phys. Letters* *3*, 666 (1969).

# CATION RADICAL INTERMEDIATES, EXCIPLEXES AND

# ENCOUNTER COMPLEXES IN PHOTOCHEMICALLY-INDUCED

# POLYMERIZATION AND CYCLODIMERIZATION OF OLEFINS

A. LEDWITH

*Donnan Laboratories, University of Liverpool*

*Liverpool, L69 3BX*

## INTRODUCTION

Photochemically-induced electron-transfer reactions between neutral organic molecules were initially recognized, and frequently specified, as arising solely from photoactivation of so-called charge-transfer complexes. The latter concept arose directly from the Mulliken description (1) of the bonding and, more specifically, the electronic transitions observed to occur for combinations of donor molecules (D) - having low ionization potentials ($I_D$), and acceptor molecules (A) - having high electron affinities ($E_A$). A useful generalization arising from the Mulliken theory is that the energy of the charge-transfer transitions between organic molecules in solution is given by

$$h\nu_{CT} = I_D - E_A + C$$

where C is a constant for the particular combination of D and A and represents mainly coulombic forces. It follows,

therefore, that the excited states of charge-transfer complexes have structures resulting from a considerable degree
of electron transfer in comparison to their ground states,
*e.g.*

$$D + A \rightleftharpoons (D, A) \xrightarrow{\;h\nu\;} (D^{\overset{+}{\bullet}}, A^{\overset{-}{\bullet}})^{*}$$

Nevertheless, the excited states of a charge-transfer complex
should not be equated with a thermally-equilibrated pair of
ion radicals, a point which was first highlighted in the
early, but stimulating, review by Kosower (2).

More recently, mainly as a result of the pioneering
studies of Weller and his associates (3), it has become
apparent that excited states, having structures equivalent to
those of photoexcited charge-transfer complexes, may be
formed by local excitation of one or the other component
(*i.e.* D or A) in systems which do not give evidence of ground-
state complex formation, *e.g.*

$$D \xrightarrow{\;h\nu_D\;} D^{*} \xrightarrow{\;A\;} (D^{*}, A) \longrightarrow (D^{\overset{+}{\bullet}}, A^{\overset{-}{\bullet}})^{*}$$

$$A \xrightarrow{\;h\nu_A\;} A^{*} \xrightarrow{\;D\;} (D, A^{*}) \;\; \underline{\qquad\qquad}$$

Here the excited state having the most electron-transfer
character $(D^{\overset{+}{\bullet}}, A^{\overset{-}{\bullet}})^{*}$ is termed an exciplex and its formation
may be preceded by a variety of collisional complexes
(encounter complexes) between an excited donor and ground-
state acceptor or vice versa, *e.g.* $(D^{*}, A)$ and $(D, A^{*})$.

Characterization of the photophysical and some photochemical consequences of exciplex formation are the subjects
of other papers in this book and, additionally, the reader
is referred to the review by Ottolenghi (4) for an informa-
tive account of the interrelationships between excited states
of charge-transfer complexes and ion pairs derived from

locally excited states of either donor or acceptor component.

Whether or not we are dealing with photoexcitation of components interacting in their ground states, or those involving a locally excited state-ground state combination, there is an immediate bonus to the practical chemist in that both afford highly convenient ways of generating ion radical intermediates from neutral organic molecules. It will be the purpose of this survey to indicate how such interactions, and excited states, may prove to be of technological value in the field of polymer chemistry and, in the process, have led to the characterization of a new type of cycloaddition reaction.

## PHOTOINITIATION OF FREE RADICAL POLYMERIZATION

The various ways by which a photoexcited molecule can initiate vinyl polymerization have been outlined elsewhere (5, 6) and, for present purposes, can be restricted to those involving aromatic carbonyl compounds. Three types of photochemical reaction may be involved:

(i) Photofragmentation of molecules such as benzoin and its alkyl ethers, $e.g.$

$$C_6H_5 - \overset{\overset{\text{O}}{\|}}{C} - \overset{\overset{\text{OR}}{|}}{CH} - C_6H_5 \xrightarrow{h\nu} C_6H_5\overset{.}{C}O + \cdot CH - C_6H_5$$

(ii) Photoinduced hydrogen transfer from substrate, solvent etc. to the excited carbonyl compound, $e.g.$

$$(C_6H_5)_2C = O \xrightarrow{h\nu} [(C_6H_5)_2 C = O]^* \xrightarrow{RH} (C_6H_5)_2\overset{.}{C} - OH + R\cdot$$

(iii) Photoinduced electron transfer from substrate, solvent etc., to the excited carbonyl compound, $e.g.$

$$Ar_2 C = O \xrightarrow{h\nu} (Ar_2 C = O)^* \xrightarrow{R_3N} (Ar_2 C = O)^{\overline{\cdot}} + R_3 N^{\overset{+}{\cdot}}$$

From a technological point of view, photodecomposition of benzoin derivatives is the most useful of these processes at the present time because of availability, high quantum efficiency for photodecomposition and high initiating efficiency of the resulting radical fragments (7). Hydrogen abstraction reactions by photoexcited benzophenone derivatives have high quantum efficiencies but have found little use in photoinitiation of vinyl polymerization perhaps because of complicating side reactions, including termination, induced by the attendant semipinacol radicals (e.g. $(C_6H_5)_2 \overset{\cdot}{C} - OH$) (8, 9). In contrast to reactions under (i) and (ii) which involve mainly $n,\pi^*$ excited states of aromatic carbonyl compounds, those under (iii) are equally efficient for $\pi,\pi^*$ excited states; their characterization is a result of comparatively recent work by the groups led by Cohen in the United States and by Davidson in England. A full account of photoreduction of aromatic ketones by amines has been given by Cohen, Parola and Parsons (10) and it can now be seen that these reactions constitute just one class of a wider range of photoreactions of amino compounds occurring via formation of exciplexes (3, 11). Denoting aromatic carbonyl compound or condensed aromatic molecule, etc. by A, photooxidation of amines may be generalized as

$$A \xrightarrow{h\nu} A^* \xrightarrow{R_3N} (R_3N, A^*) \longrightarrow (R_3N^{\overset{+}{\cdot}}, A^{\overline{\cdot}})^* \longrightarrow \text{products}$$

Of particular interest here are the processes whereby encounter complexes $(R_3N, A^*)$ and exciplexes $(R_3N^{\overset{+}{\cdot}}, A^{\overline{\cdot}})^*$ may give rise to formation of free radicals active in the initiation of vinyl polymerizations (12).

212

Singlet and triplet states of aromatic ketones are efficiently quenched by a wide variety of amino compounds and a
general correlation exists between quenching efficiency
and ionization potential of the amine, for a particular class
of amine (10). In many cases tertiary amines are more
effective than the corresponding secondary and primary derivatives although specific solvation phenomena may change this
order of reactivity. Benzophenone, no less than other aromatic ketones, is efficiently reduced by amines in a variety of
solvents at rates which are substantially faster than those
observed for corresponding photoinduced hydrogen abstractions
from, say, alcohols (13). Products from these various photoredox processes are related and, for the photooxidation of
amines, may be generalized as follows:

$$Ar_2C=O \xrightarrow{h\nu} (Ar_2C=O)^* \xrightarrow{RCH_2CH_2NR_2} (Ar_2C=O)^{\cdot -} (RCH_2CH_2NR_2)^{\cdot +}$$

$$\longrightarrow Ar_2\dot{C}\text{-OH} + RCH_2\dot{C}HNR_2$$

$$Ar_2\dot{C}OH + RCH_2\dot{C}HNR_2 \longrightarrow Ar_2CHOH + RCH=CHNR_2$$

$$RCH_2\dot{C}HNR_2 + Ar_2C=O \longrightarrow RCH=CHNR_2 + Ar_2\dot{C}\text{-OH}$$

$$2\ Ar_2\dot{C}OH \longrightarrow \underset{\underset{OH}{|}}{Ar_2C} - \underset{\underset{OH}{|}}{C}\ Ar_2$$

In our own work we have preferred to utilize ketones
thought to have lowest lying $\pi,\pi^*$ triplet excited states
because, although the point is still uncentain in some cases,
$\sigma,\pi^*$ excited triplets do not generally undergo direct hydrogen abstraction reactions with alcohols, alkanes, etc.
Fluorenone (FLO) is undoubtedly the most useful of the aromatic ketones having lowest lying $\pi,\pi^*$ triplet excited states
and its photophysical characteristics have been extensively

studied by several groups of workers (14 - 16). N–Methylacridone (NMA) is also useful because of its greater light
absorption characteristics.

Photoreduction of fluorenone does not occur in alcohol,
ether, or alkane solvents but occurs readily in the presence
of amines, with tertiary amines being most effective (17, 18).
A rather special feature of the photochemistry of fluorenone
is the dramatic effect of solvent on the rates of intersystem
crossing from singlet to triplet manifolds (15, 16, 19).
Although a detailed explanation of this phenomenon is still
lacking, the generalization may be made that increasing solvent polarity decreases the facility for intersystem crossing
and simultaneously increases the quantum yield for fluorescence. Thus, triplet-state activity is maximized in solvents
such as benzene and cyclohexane, for which the values of triplet yields are 0.93 and 1.03 respectively (19). In contrast,
triplet yields in solvents such as alcohols, acetone and
acetonitrile are substantially less than unity with a concomi-
tant increase in quantum yields for fluorescence (15, 16).
Ionization potentials of amines are important in determining
quenching ability for both singlet and triplet excited
fluorenone but it now appears that *photoreduction* is consequent upon interaction of triplet excited fluorenone with
amine donors (10), the main effect of singlet-state quenching
being to reduce triplet yields. Experimental observations
from several groups of workers may be summarized, therefore,
by stating the the photoreduction of fluorenone by amines
occurs most readily in solvents such as benzene and cyclohexane and only with low quantum efficiency in solvents such
as acetone and acetonitrile. A further complication arises
because the amines themselves may be regarded as polar additives - in comparison to hydrocarbons - and hence there is a

marked effect of amine concentration on photoreduction of
fluorenone in hydrocarbon solvents. For example, quantum
yields for photoreduction of fluorenone by triethylamine have
been measured to be 0.09 in neat amine, 0.9 in 0.1 $M$ amine
(in cyclohexane) and 0.6 in 0.02 $M$ amine (10). That electron
transfer is involved in these quenching and photoreduction
processes is indicated not only by the clear dependence on
ionization potential of the donor amine, but also by flash
photolytic and ESR characterization of ion radical intermediates, especially for those systems where the quencher has
no readily abstractable hydrogen atom (*e.g.* triarylamines) (20).

We have observed that the photoreduction of fluorenone by
suitable tertiary amines affords a highly convenient photoinitiation system for free radical polymerization (21).
Almost any tertiary amine may be used but, for experimental
convenience, N,N-dimethylethanolamine (DME) has been the most
studied. Even greater efficiencies of radical formation
(per mole of amine donor) were observed for comparable reduction using indole-3-ylacetic acid (IAA) as donor, although
here the overall reaction mechanism is thought to involve
photodecarboxylation (22) and will not be discussed further.

FLO    NMA    DME

IAA    MMA

As a model system, photoinitiation of polymerization of
methyl methacrylate (MMA) was fully studied in solvents
benzene, acetone, acetonitrile) by conventional dilatometric

techniques in high vacuum systems.

Experimental data for the variations in polymerization reaction rate ($R_p$) with sensitizers (FLO) are given in figure 1.

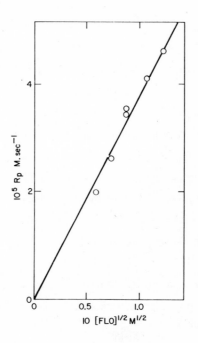

Fig. 1    *Polymerisation of MMA in benzene at $30°C$.    DME/ FLO system.    Variation of $R_p$ with [FLO].*

$R_p$ was found to be proportional to the half power of the ligh intensity ($I_o$) and, for a fixed amine concentration, the over all kinetic expression best fitting the experimental data was

$$R_p = \frac{-d[MMA]}{dt} = k_p [FLO]^{\frac{1}{2}} [I_o]^{\frac{1}{2}} [MMA]$$

The effect of varying [DME] on polymerization reaction rate was more complex.   As indicated in figure 2, $R_p$ appears to pass through a maximum before becoming essentially constant

as [DME] is increased. This result is rather similar to
those reported for effects of increasing amine concentrations
on quantum yields for photoreduction of fluorenone (*vide
supra*).

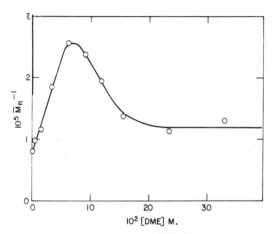

*Fig. 2  Polymerisation of MMA in benzene at 30$^{o}$C.
DME/FLO system.  Variation of $R_p$ with [DME].*

It is important to note (see figure 3) that the maximum in
rate of polymerization as a function of increasing [DME] is
matched by a corresponding minimum in the number average
molecular weight ($\overline{M}_n$).  Indeed evaluation of the ratio $k_p/k_t^{\frac{1}{2}}$
for the system DME/FLO gave a value of 0.054 (30$^{o}$C) which is
in very good arreement with the value 0.057 (25$^{o}$C) reported
by Bamford and Brumby (23) for initiation of MMA polymerization in the same solvent (benzene) by azobisisobutyronitrile.
It follows therefore that transfer and termination processes
in the DME/FLO system are not particularly affected by the
initiating components or their photoredox products.  A
further point of similarity between the photoinitiation of
free radical polymerization and previously established

217

features of the photoreduction of fluorenone by amines, is
to be found in the effect of polar solvents.

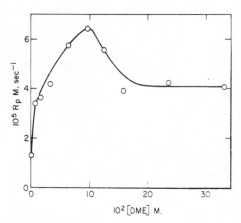

Fig. 3   Polymerisation of MMA in benzene at $30^{\circ}C$.
DME/FLO system.   Variation of $\overline{M}_n^{-1}$ with [DME].

TABLE 1

Effect of Solvent on Polymerization of MMA at $30^{\circ}$ C

| Solvent | $R_p$ ($M$ sec$^{-1}$) |
|---------|------------------------|
| Benzene | $2.35 \times 10^{-5}$ |
| Acetone | $1.03 \times 10^{-5}$ |
| Acetonitrile | $0.50 \times 10^{-5}$ |

[DME] $= 6.28 \times 10^{-4}$ $M$, [FLO] $= 1.13 \times 10^{-4}$ $M$

[MMA] $= 2.00$ $M$, $\lambda = 366$ nm

Table 1 shows that $R_p$ decreases for standard polymerization
runs as the solvent changes from benzene to acetonitrile, in
complete accord with the idea that triplet excited fluorenone
is responsible for initiation.   Independent determinations of
quantum yields for photoreduction of fluorenone by the

particular amine in question (DME) show an identical effect
of solvent polarity (table 2) and are in substantial agreement with related data of other workers (10). However the
data of table 2 indicate an additional important reaction
involving apparent quenching of triplet excited fluorenone by
the polymerizable monomer MMA. This effect is quite real and
reproducible and reaches a saturation value at approximately
1 $M$ olefin (table 3). An exact value of the triplet-state
energy of MMA is not available but it must be considerably
greater than that of fluorenone ($E_T$ = 53 kcal).

TABLE 2

*Quantum Yields ($\phi$) for Photoreduction of Fluorenone (FLO)*

*by $(CH_3)_2NCH_2CH_2OH$ (DME)*

| Solvent | [FLO] $10^3$ $M$ | [DME] $10^3$ $M$ | $\phi$ |
|---|---|---|---|
| Benzene | 2.0 | 2.0 | 1.20 |
| Benzene | 2.0 | 2.0 | 1.23 |
| Benzene/MMA | 2.0 | 2.0 | 0.35 |
| 2:1 (v/v) | 2.0 | 2.0 | 0.38 |
| $CH_3CN$ | 2.0 | 2.0 | 0.36 |
| $CH_3CN$/MMA | 2.0 | 2.0 | 0.42 |
| 2:1 (v/v) | | | |

(In vacuo, $\lambda$ = 366 nm, temp. $30^{\circ}$ C)

Quenching of triplet fluorenone by the normal electron
exchange mechanism can, therefore, be eliminated as a possibility and it would appear that quenching by MMA is another
example of the (as yet ill-defined) energy wastage mechanisms

TABLE 3

*Effect of MMA on Quantum Yield ($\phi$) of Photoreduction of*

*Fluorenone (FLO) by $(CH_3)_2NCH_2CH_2OH$ in Benzene at $30^{o}C$*

| [MMA] | $\phi$ |
|-------|--------|
| 0 | 1.20 |
| 0.089 | 0.55 |
| 0.89 | 0.36 |
| 3.46 | 0.35 |

[DME] = [FLO] = $2.0 \times 10^{-3}$ M
(in vacuo)

for triplet excited fluorenone previously noted by Caldwell and Gajewski (19) and Valentine and Hammond (24). Energy wastage could involve (a) chemical reaction of triplet excited fluorenone with ground-state MMA to give an unstable (biradical) adduct which reverts to MMA and ground-state fluorenone or alternatively (b), a weakly bound encounter complex between triplet fluorenone and MMA might provide the necessary vibrational pathway for overall deexcitation, *e.g.*

(a) $FLO^* + CH_2=C-COOCH_3 \longrightarrow \cdot FLO-CH_2-\overset{\cdot}{C}-COOCH_3 \longrightarrow FLO + MMA$
$\phantom{(a) FLO^* + CH_2=}\underset{CH_3}{|}\phantom{-COOCH_3 \longrightarrow \cdot FLO-CH_2-}\underset{CH_3}{|}$

(b) $FLO^* + MMA \longrightarrow (FLO^*...MMA) \longrightarrow FLO + MMA$

It is also conceivable that the effect of MMA represents a further manifestation of the dramatic effects of medium on triplet yields for fluorenone.

A simplified general expression for rates of polymerization is

$$R_p = \frac{k_p}{k_t^{\frac{1}{2}}} [M] (R_i)^{\frac{1}{2}}$$

where $R_i$ is the rate of initiation and includes terms for the quantum yield for production of initiating radicals, light absorption characterisitcs of the sensitizer and incident light intensity. Knowing $R_p$ and $k_p/k_t^{\frac{1}{2}}$, it is possible to estimate values of the quantum yields for *initiation* of polymerization of MMA and these are shown in table 4.

TABLE 4

*Spectroscopic Data and Quantum Yields for Initiation, $\phi$, of MMA Polymerization in Benzene at $30^o$ C*

| Photoinitiator | $\lambda_{max}$ (nm) $\varepsilon_{max}$ (1 mol$^{-1}$ cm$^{-1}$) | Amino com- pound | $\lambda$ (nm) for measurement of $\phi$ | $\phi$ |
|---|---|---|---|---|
| 9-Fluorenone | 380 (321) | DME | 366 | 0.22 |
| 9-Fluorenone | 380 (321) | IAA | 366 | 0.06 |
| 9-Fluorenone | 380 (321) | N-Me-IAA | 370 | 0.21 |
| N-Methylacridone | 460 (8470) | IAA | 406 | 0.07 |

Taking the data for DME/FLO systems, quantum yields for photoinitiation of free radical polymerization of MMA are of the same order as those for photoreduction of fluorenone under comparable conditions. The difference ($\phi_{red}$ = 0.35, $\phi_{init}$ = 0.2) may be more apparent than real especially in view of the uncertainties and approximations associated with the estimation of quantum yields for initiation. Two other

effects are evident from the data in table 4; (i), the
greatly increased light absorption characteristics of N-
methylacridone over fluorenone offer potential improvements
in overall efficiency of initiation and (ii), quantum yields
for initiation are significantly higher for FLO/N-methyl-
indole-3-ylacetic acid than FLO/IAA systems. Conceivably the
N-H bond in IAA provides an additional energy wastage
mechanism for the excited ketone, as found for photoreduction
of fluorenone by secondary amines (10, 25).

The reaction sequence accounting for all the experimental
observations for both photoreduction of fluorenone and
photoinitiated polymerization is given in scheme 1.

Scheme 1

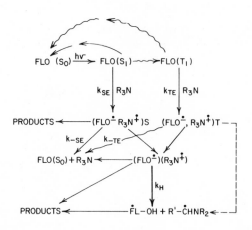

Here the quenching and/or energy wastage mechanisms for
excited fluorenone not involving amine are represented, for
clarity, by the curved arrows at the top of the diagram.
Singlet (S) and triplet (T) exciplexes ($FLO^-$, $R_3N^+$) are produced from the appropriate excited states by collisional

encounters with amine molecules ($k_{SE}$ and $k_{TE}$ respectively) and may undergo radiationless decay to ground states by the reverse processes ($k_{-SE}$, $k_{-TE}$). Chemical products are seen as arising directly from the triplet exciplex or, *via* a pair of ion radicals $(FLO^{-})(R_3N^{+})$, formed by further relaxation of either singlet or triplet exciplexes, between which the proton transfer ($k_H$) is kinetically important. Solvent effects previously noted suggest that, for production of radical species, relaxation of singlet exciplexes $(FLO^{-} R_3N^{+})S$, does not compete as favorably with $k_{-SE}$ as does the comparable relaxation of triplet exciplexes $(FLO^{-}, R_3N^{+})T$ with $k_{-TE}$ , or direct formation of the *radicals* from the triplet exciplexes. Increasing the solvent polarity decreases the efficiency of the conversion, $FLO(S_1) \longrightarrow FLO(T_1)$, and also provides for a radiationless deactivation pathway for $FLO(S_1)$.

Independent studies (9, 26) indicate that FL-OH and related semipinacol radicals are not especially efficient in initiation of polymerization at ambient temperatures and the important initiating species will, therefore, be amine derived radicals such as $R\dot{C}HNR_2$. FLO/DME systems have been used successfully to initiate the polymerization of methyl methacrylate, methyl acrylate, acrylonitrile and styrene; rather similar mechanisms may be involved in the photoinitiation of methyl acrylate by benzophenone/triethylamine (27) and the recently reported photoinitiated polymerization of methyl methacrylate by 4,4'-bis(diethylamino)benzophenone (28).

Technological applications of photoinitiated polymerizations are discussed in the final part of this review, but it is worth noting at this juncture that photoinitiation *via* exciplexes such as are formed between fluorenone and tertiary amines affords a highly convenient method for synthesis of block and graft copolymers, as indicated in scheme 2. For block polymerization, the preformed polymer from one FLO-amine initiated polymerization, which will necessarily possess at least one amine end group, can be used as the amino component of a second polymerization. For graft copolymerization, it is convenient to use preformed polymers produced by copolymerization of suitable monomers with the readily available 2-(N,N-dimethylamino)ethyl methacrylate,

$$(CH_3)_2NCH_2CH_2OOCC=CH_2.$$
$$\underset{CH_3}{|}$$

Scheme 2

BLOCK COPOLYMERISATION

$$FLO + (CH_3)_2 NR \xrightarrow{h\nu} \overset{\bullet}{F}LOH + \overset{\bullet}{C}H_2\overset{R}{\underset{|}{N}}-CH_3$$

$$CH_3\overset{R}{\underset{|}{N}}-CH_2\bullet + M_1 \longrightarrow CH_3\overset{R}{\underset{|}{N}}CH_2M_1\sim\sim\sim\sim M_1\bullet \text{ etc.}$$

$$CH_3\overset{R}{\underset{|}{N}}CH_2M_1\sim\sim\sim\sim\sim\sim M_1X$$

$$M_2 \mid h\nu \text{ (FLO)}$$

$$etc \quad \bullet M_2\sim\sim\sim\sim M_2CH_2\overset{R}{\underset{|}{N}}CH_2M_1\sim\sim\sim\sim M,X$$

GRAFT COPOLYMERISATION

$$\sim\sim M_1\sim\sim\sim\sim\sim\sim\sim M_1\sim\sim\sim$$
$$\underset{N(CH_3)_2}{|} \qquad \underset{N(CH_3)_2}{|}$$

$$M_2 \mid h\nu \text{(FLO)}$$

$$\sim\sim M_1\sim\sim\sim\sim\sim\sim M_1\sim\sim\sim$$
$$\underset{CH_3N}{|} \qquad \underset{CH_3N}{|}$$
$$\underset{CH_2M_2\sim}{} \qquad \underset{CH_2M_2\sim}{}$$
$$M_2\bullet \qquad M_2\bullet$$

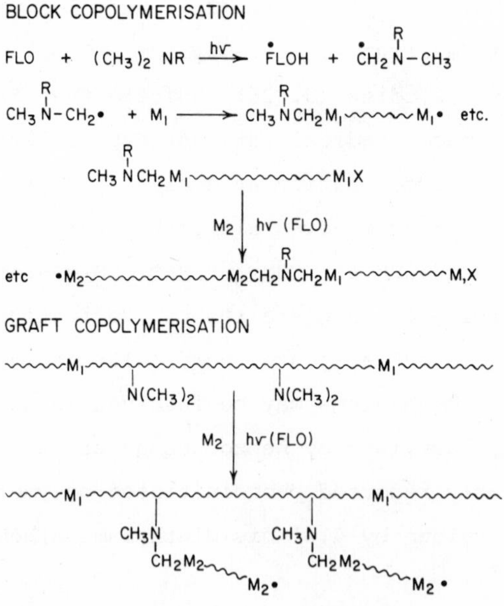

Finally, while the experimental observations referred to above were made by studies on vacuum degassed systems, similar reactions may be carried out quite conveniently in air. In the latter case, small induction periods are observed due to the scavenging of oxygen by radical products of the photoredox processes such as $R\dot{C}HNR_2$ and $\dot{F}L\text{-}OH$.

PHOTOINDUCED CYCLODIMERIZATION OF N-VINYLCARBAZOLE

Dimerization of N-vinylcarbazole (NVC) to the cycloadduct *trans* - 1,2-dicarbazylcyclobutane (DCB), is readily accomplished by a variety of electrochemical (anodic) (29), thermal (metal oxidant) (30), photochemical (31) and radiolysis (32) techniques. In many cases, the cyclodimerization is accompanied by free radical or cationic polymerization of NVC to give poly(N-vinylcarbazole). A general feature of all these catalytic processes is an initial requirement for removal of one electron to form the monomer cation radical (NVC)$^{+}$, hence the wide range of oxidation techniques which are effective (33).

The photoinduced process is of primary interest and was first characterized by Ledwith *et al.* (31, 33) who showed that several olefins, having structures and oxidation potentials rather similar to NVC, could be effectively cyclodimerized by photolysis in the presence of a wide range of sensitizers in methanol or acetone solutions.

NVC

A striking feature of these photoinduced cyclodimerizations was the requirement, in the case of many sensitizers, for the presence of dissolved air or oxygen. It is now clear that molecular oxygen, or other suitable thermal oxidants (*e.g.* copper(II) heptanoate), plays a cocatalytic role especially for the long wavelength dye sensitizers. Significantly , the sensitizers which function in the absence of oxygen are those such as quinones and fluorenone which yield comparatively stable anion radicals. In all cases, however, the presence of oxygen increases the apparent rate of photo-dimerization. An indication of the range of sensitizers which may be employed, together with their triplet-state energies and the requirements for a cooxidant (usually $O_2$) is given in table 5.

TABLE 5

*Photosensitizers for Cyclodimerization of N-Vinylcarbazole in Methanol*

| Sensitizer | $E_T$ (kcal) | Cooxidant Required |
|---|---|---|
| Xanthone | 74 | Yes |
| Benzophenone | 69 | Yes |
| Anthraquinone | 62 | No |
| Chloranil | 57 | No |
| Fluorenone | 53 | No |
| 2,4,7-Trinitrofluorenone |  | No |
| 2,4,6-Triphenylpyrylium$^+$BF$_4^-$ | 53 | Yes |
| Acriflavine | 51 | Yes |
| Dibromo (R) fluorescein | 47 | Yes |
| Rhodamine 6 G | 43 | Yes |
| Methylene blue | 33 | Yes |
| (Carbazole) | (70) | (Yes) |

It is immediately apparent that the normal electron-exchange mechanism of energy transfer from triplet excited sensitizer to NVC cannot account for most of the photocatalyzed reactions because of the comparatively high values of triplet energies in carbazole derivatives. By use of appropriate wavelength filters, it was shown that the sensitizer was the primary light absorbing species in all cases and the cyclo-dimerizations were essentially quantitative if irradiation was prolonged.

Quantum yields for photosensitized cyclodimerization of NVC were measured under a variety of experimental conditions, for different sensitizers, and are given in table 6.

TABLE 6

*Quantum Yields for Cyclodimerization of NVC*

| Sensitiser | Solvent | Atmosphere | $\phi$ |
|---|---|---|---|
| Benzophenone | Acetone | Air | 0.8 |
| " | MeOH | " | 1.0 |
| Tetramethyloxycarbonyl-benzophenone | Acetone | " | 3.3 |
| Decafluorobenzophenone | Acetone | Air | 66 |
| " | " | $N_2$ | 55 |
| Fluorenone | Acetone | Air | 3.6 |
| " | " | $N_2$ | 2.1 |
| 2,4,7-Trinitro-fluorenone | Acetone | Air | 7.6 |
| " | " | $N_2$ | 4.9 |
| " | MeOH | Air | 1.0 |
| Chloranil | Acetone | Air | 8.5 |
| " | " | $N_2$ | 4.3 |
| " | MeOH | Air | 2.9 |
| " | " | $N_2$ | 1.1 |

[NVC] = $10^{-1}$ *M*, [Sensitiser] = $10^{-2}$ -$10^{-3}$ *M*
Irradiation with $\lambda$ = 366 nm

The data of table 6 show the relative cocatalytic effects of dissolved oxygen but provide unequivocal evidence that these photoinduced cyclodimerizations are chain reactions; quantum yields are usually greater than unity and reach values of ~ 60 for sensitization by decafluorobenzophenone. A further point of interest is the correlation between the quenching of the sensitizer fluorescence and the apparent efficiency in sensitization of cyclodimerization. Thus NVC is an effective quencher for the fluorescence of fluorenone (in acetone) and rhodamine 6G (in methanol), which are taken as representative examples of the wide range of useful sensitizers. Additives such as ferrocene, DABCO (diazobicyclooctane), and iodide ion, which are very good electron donors, generally quench the fluorescence of the sensitizer and retard the cyclodimerization (table 7).

TABLE 7

*Retardation of Cyclodimerisation of NVC and Fluorescence Quenching*

| Quencher | Fluorenone in Acetone | | Rhodamine 6G in Methanol | |
|---|---|---|---|---|
| | $K_Q(M^{-1})$ | Rel. Rate of Dimerisation | $K_Q(M^{-1})$ | Rel. Rate of Dimerisation |
| NVC | 88 | 1.00 | 23.0 | 1.00 |
| DABCO | 171 | $0.10^a$ | 8.0 | $1.00^a$ |
| Ferrocene | 268 | $0.00^a$ | 88.1 | $0.04^a$ |
| $I^-$ | 395 | $0.15^a$ | 90.2 | $0.16^a$ |
| $Br^-$ | | | 0 | $1.00^a$ |
| $Cl^-$ | | | 0 | $1.00^a$ |
| EtI | | | 0 | $1.00^a$ |

Temp., 25°C

[a] Reaction mixture contained $10^{-3}M$ (Quencher)

It is significant that DABCO, a more efficient quencher of the fluorescence from fluorenone than is NVC in acetone, retards the photodimerization of the latter, whereas with rhodamine 6G in methanol, in which DABCO is a less efficient quencher than NVC, there is no effect on the rate of photo-dimerization. A chain reaction mechanism (34) accounting for most of the experimental observations is indicated in scheme 3. The initiation reaction applies equally well to singlet or triplet excited states although the former are thought to be primarily involved (*vide infra*).

### Scheme 3

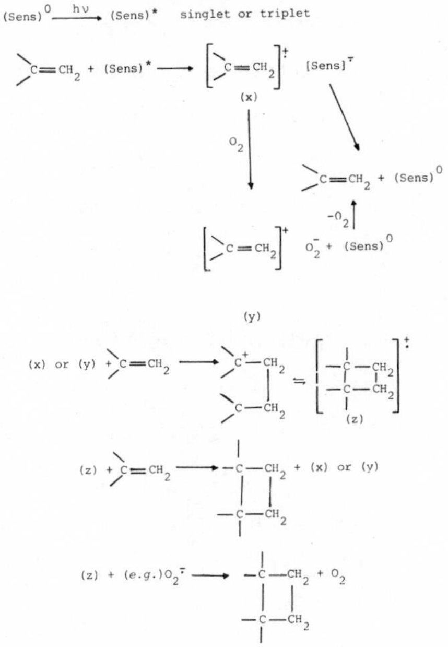

Apparent quantum yields observed for any reaction will depend on the relative values of the rate constants indicated in the scheme which, in turn, will depend upon the nature of the sensitizer, its reduction potential and the interactions of all intermediates with solvents and with oxygen.

In the absence of oxygen several of the sensitizers bring about (mainly free radical) photoinduced polymerizations of

NVC rather than cyclodimerization. It is generally the case, however, that the quantum yields for the *initiation* of radical polymerization of NVC are orders of magnitude lower than those for the corresponding cyclodimerizations on admission of air to the systems. Accordingly, observation of polymer formation in systems exhibiting charge-transfer complex, or exciplex, behavior may not represent a particularly efficient photoreaction. The interrelationship between cyclodimerization and polymerization of NVC as a result of photolysis in the presence of electron acceptors, has been extensively investigated by Mikawa and collaborators (35, 36). In particular, these workers have shown that, for a given sensitizer, cyclodimerization is favored by more basic solvents such as acetone, methanol and acetonitrile, while less basic solvents such as benzene favor polymerization. The NVC cation radical (37) is known to be a primary intermediate in all these photoprocesses and it is at least possible that radical and cationic polymerizations would be initiated. However more detailed studies of the quantum yields for individual reaction types and of the role of oxygen (cocatalyst or inhibitor?) in the two types of solvent are required before the photopolymerization processes can be properly defined. The range of olefins which may be photodimerized by the cation radical chain reaction is very limited (30) but, in highly significant developments, has recently been extended by Farid and Shealer (38) to include indene and derivatives, and by Evans and coworkers (39) to phenyl vinyl ether.

Reactions of N-vinylcarbazole, a tertiary amine, photosensitized by fluorenone, will now be considered in more detail and compared with those discussed in the early part of this survey.

Previously we have noted that photoinitiation of free

radical polymerization by tertiary amine-fluorenone systems is favored by solvents of low polarity (*e.g.* benzene) and involves ultimate electron transfer between amine and *triplet* excited fluorenone. The parallelism between fluorescence quenching and photoinduced cyclodimerization of NVC argues strongly for a mechanism involving ultimate electron transfer between NVC (the tertiary amine in this case) and *singlet* excited fluorenone. Furthermore, the solvent effect is just the opposite of that noted for photoinitiation of free radical polymerization (table 8). Quite clearly, the nonpolar solvents, benzene and cyclohexane, do not help promote formation of cyclodimer. It should be noted here that while methanol is an excellent solvent for use with other sensitizers, it is less good than acetone for fluorenone-sensitized reactions largely because methanol (and other protic solvents) are weak quenchers of the fluorescence of fluorenone. Typical values of the Stern-Volmer quenching constants $K_Q (M^{-1})$ for fluorenone in acetone are: $CH_3OH$ (1.25), $CD_3OD$ (0.74), $H_2O$ (0.53), $D_2O$ (0.27).

TABLE 8

*Effect of Solvent on Cyclodimerization of NVC* **Sensitized** *by Fluorenone in Air*

| Solvent | Reaction Time (Hrs.) | Product | Yield % |
|---------|----------------------|---------|---------|
| Methanol | 5.0 | Dimer | 12 |
| Acetone | 0.5 | Dimer | 62 |
| Benzene | 90.0 | Polymer | 7 |
| Cyclohexane | 90.0 | – | 0 |

$[NVC] = 0.1\ M$,    $[FLO] = 10^{-2}\ M$

Final proof that singlet excited fluorenone is the important photoreactant in these systems comes from inspection of appropriate polarographic redox potentials. The half-wave reduction potential (vs S.C.E.) for fluorenone in dimethylformamide (40) is reported to be -1.29 $V$; the corresponding potential for oxidation of NVC (vs S.C.E.) in acetonitrile (41) is approximately +1.2 $V$. It follows that generation of the solvent-equilibrated ion pair (NVC$^{+}$) (FLO$^{-}$) requires at least ~ 2.5 $V$ or ~ 60 kcal mole$^{-1}$. Exciplexes would be expected to have higher energies than solvent-equilibrated ion pairs, depending on the solvation energies; hence it can be concluded that triplet fluorenone ($E_T$ = 53 kcal) is unlikely to be involved in overall electron transfer reactions of carbazole derivatives. Aliphatic amines, and other aromatic amines useful in photoinitiation of free radical polymerization, have half-wave oxidation potentials some 0.3 - 0.5 $V$ lower than these of carbazole compounds and hence could be oxidized by triplet excited fluorenone (42).

Referring now to scheme 1, and in contrast to the photoinitiation processes, photoinduced cyclodimerization of NVC by fluorenone must involve product formation (*i.e.* cyclobutane) directly from the singlet-state exciplex (FLO$^{-}$, R$_3$N$^{+}$)S or, much more probably, from the subsequently derived pair of ion radicals (FLO$^{-}$)(R$_3$N$^{+}$). Evidence in favor of the latter possibility is provided by the relative reactivities, in cyclodimerization, of *cis*- and *trans*-N-propenylcarbazole (scheme 4).

Whereas trans-N-propenylcarbazole may be cyclodimerized by various photosensitized and metal-catalyzed processes useful for the simple N-vinyl compound, *cis*-N-propenylcarbazole is completely unreactive and may be recovered unchanged from attempted photosensitized cyclodimerizations. On the

Scheme 4

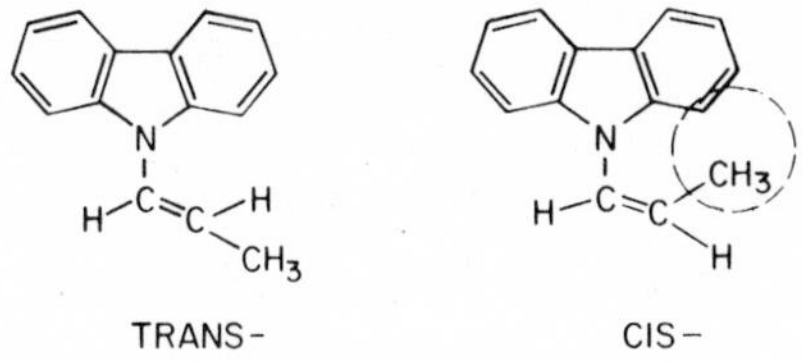

TRANS-                    CIS-

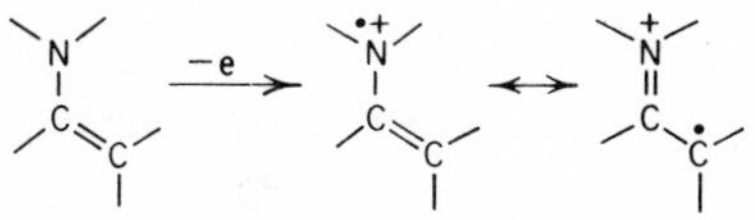

other hand, ionization potentials and electron donor
quenching ability towards rhodamine 6G are very similar for
both *cis*- and *trans*-N-propenylcarbazoles and NVC (table 9).

TABLE 9

*Fluorescence Quenching of Rhodamine 6G in Methanol*

| Quencher | $K_Q (M^{-1})$ | Vertical IP (eV) (P.E. Spectra) |
|---|---|---|
| N-Vinylcarbazole | 23.0 | 7.42 |
| N-Ethylcarbazole | 40.0 | 7.29 |
| *cis*-N-Propenylcarbazole | 40.0 | 7.26 |
| *trans*-N-Propenylcarbazole | 51.2 | 7.12 |
| Temp., 25$^\circ$C | | |

It may be concluded, therefore, that the ability of a
carbazole moiety to give up an electron is not significantly
impaired by vinyl or propenyl substitution on nitrogen.

However, inspection of molecular models shows clearly that, because of excessive steric interaction of the *cis-* methyl group and the 1,8-ring positions, *cis*-N-propenylcarbazole cannot adopt a conformation in which the unshared electron pair on nitrogen and the olefinic π-orbitals are coplanar. Such coplanarity is a necessary requirement for resonance delocalization of the carbazole cation radical into the olefinic group (scheme 4), and this, in turn, is a requirement for reaction of the cation radical with neutral olefin to generate the chain reaction leading to cyclobutane dimer. It is reasonable to conclude, therefore, that the decay of the singlet exciplex for fluorenone and *cis*-N-propenylcarbazole (*i.e.* $k_{-SE}$ in scheme 1) is more rapid than the subsequent formation of the equivalent pair of ion radicals or of cyclodimer product. Isomerization of *cis-* to *trans*-N-propenylcarbazole is not observed during quenching of the fluorescence of the sensitizers and this also argues strongly in favor of quenching *via* a singlet-state exciplex which can return to ground-state components more rapidly than it can relax to a pair of ion radicals. It is interesting to speculate that delocalization of the cation radical (*cf* scheme 4) in say, NVC, favors the relaxation of FLO-NVC singlet exciplexes to form reacting ion pairs or, just possibly, facilitates direct reaction of singlet exciplexes with ground-state NVC to initiate the chain reaction. In this way it would be possible to understand why singlet excited states of fluorenone are the important species for reacting with the tertiary amine, NVC, whereas triplet excited fluorenone appears to be more important for reaction with other tertiary amines.

CHARGE TRANSFER INTERACTION OF CHLORANIL AND NVC

The data of table 6 show that chloranil is a highly efficient sensitizer for cyclodimerization of NVC in the solvents, methanol and acetone. Chloranil, however, differs from many of the other sensitizers in that it forms charge-transfer complexes, with carbazole derivatives, including NVC. A typical charge-transfer spectrum for chloranil-NVC is shown in figure 4; for all carbazole compounds, there are two transitions, readily discernible from the asymmetric nature of the absorption curves, corresponding to charge-transfer transitions of the two highest-energy filled orbitals of the carbazole moiety (43). Photoelectron spectra of NVC indicate vertical ionization potentials of 7.42 and 7.91 e$V$, respectively, for these two states and the separation ($\sim$ 0.5 e$V$) is in excellent agreement with the energy separation of the two charge-transfer transitions with a series of acceptor molecules (44).

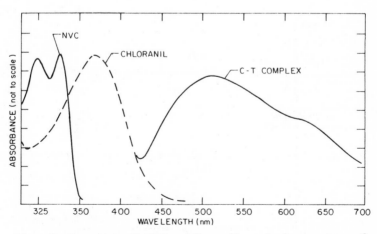

Fig. 4  Spectra of NVC, chloranil and charge-transfer complex in methylene chloride.

Cyclodimerization of NVC requires formation of the cation radical (NVC$^{+\cdot}$) which then reacts with NVC in an initiation process (scheme 3) which is readily induced by formation of exciplexes between NVC and a wide range of excited sensitizers. We have already noted that photoexcitation of ground-state charge-transfer complexes (*via* the charge-transfer band) should lead ultimately to a solvent-relaxed excited state having considerable polar character *e.g.* $(D^{+\cdot}, A^{-\cdot})^*$ and identical with a corresponding relaxed excited state formed by interactions of locally excited D or A with appropriate partners. A more detailed description of these processes is given in scheme 5.

Accordingly, it was possible to test the value of the various interaction pathways indicated in scheme 5, by photolysis of NVC-chloranil combinations in methanol, a good solvent for the cyclodimerization reaction,

Scheme 5
## CHARGE TRANSFER COMPLEX
## AND ENCOUNTER COMPLEXES

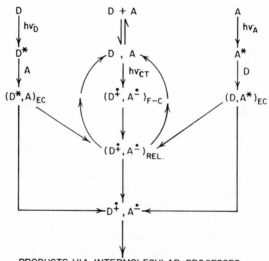

PRODUCTS VIA INTERMOLECULAR PROCESSES

Quantum yields for formation of cyclodimer are indicated in table 10; it is highly significant that selective excitation of the charge-transfer transitions (*cf* figure 5) fails to produce cyclodimer. In fact, irradiation (1 kW tungsten lamp) of 0.1 $M$ NVC and $4 \times 10^{-3}$ $M$ chloranil in methanol for twenty-four hours through a gelatin filter (max transmission at 475 nm) gave a 98% yield of N-(1-methoxyethyl)carbazole. The latter (30) is the well known methanol adduct of NVC, readily formed by acid catalysis, and represents the product of an acid-producing photochemical side reaction between methanol and chloranil, or NVC and chloranil. Thus, in spite of the anticipated ion radical nature of the excited charge-transfer states of NVC-chloranil, only local excitation of chloranil (the local excitation of NVC is also possible) yields the cyclodimer of NVC by the chain reaction mechanism.

TABLE 10

*Effect of Wavelength on Quantum Yield for Cyclodimerization of NVC by Chloranil in Air*

| $\lambda$ (nm) | $\phi$ |
|---|---|
| 366 | 4.8 |
| 405 | 2.8 |
| 546 | 0 |

[NVC] = $10^{-1} M$, [Chloranil] = $1.2 \times 10^{-2}$ $M$ in methanol

These observations provide a clear example wherein formation of a solvent-equilibrated pair of ion radicals (*e.g.* $D^{+}$, $A^{-}$ in scheme 5) is more easily accomplished *via* encounter complexes of locally excited component molecules than by exciplexes formed by excitation of ground-state charge-transfer complexes between the components. It follows,

therefore, that exciplexes or charge-transfer excited states
for the NVC-chloranil-methanol system $e.g.$ $(D^{\ddot{+}}, A^{\bar{\cdot}})^{*}_{REL}$ must
have decay pathways more efficient than equilibration with
solvent prior to initiation of the cyclodimerization reaction.
An attractive possibility is intersystem crossing of
$(D^{\ddot{+}}, A^{\bar{\cdot}})^{*}_{REL}$ to a corresponding triplet state which then
undergoes radiationless conversion to ground-state components
or, depending upon the solvent system employed, might
initiate free radical or cationic polymerization of NVC
(35, 36). It is also pertinent to recall that, since the
half-wave reduction potential of chloranil in acetonitrile is
+ 0.02 $V$ (S.C.E.) (42), generation of the ion radical pair
from an NVC-chloranil couple requires only ∼ 1.2 $V$, or 28 kcal;
hence, there is sufficient energy available $via$ any of
the charge-transfer transitions. Furthermore the individual
ion radicals have been detected by flash photolytic
techniques (37).

Other sensitizers such as 2,4,7-trinitrofluorenone and
N,N'-dimethyl-4,4'-bipyridylium dichloride, which also give
charge-transfer spectra with NVC, photosensitized the
cyclodimerization of NVC only when excited at wavelengths
shorter than could be absorbed by the appropriate charge-
transfer complexes.

TECHNOLOGICAL APPLICATIONS IN POLYMER-BASED SYSTEMS

The interaction of light with chromophores in polymer-
based systems can result in many interesting phenomena,
several of which lead to degradation or other undesirable
processes. There are, however, a growing number of technological applications where a photochemically-induced reaction
is a requisite for an improvement in the properties of the
polymer-based system. Notable among these are the wide

variety of photoinduced crosslinking reactions necessary for
improving the mechanical behavior of coatings and soft
castings based on preformed polymer systems and, more
especially, for the light-induced imaging processes required
for production of printed circuits and relief printing plates
(photoresist technology) (6). In almost all cases, the
application requires photoinduced chemical bond formation
between preformed linear polymers having suitably reactive
groupings. There are two main methods for photocrosslinking
of preformed macromolecules:

(i) Effective copolymerization of the preformed macro-
molecule, *via* in-chain or pendant functional groups (F),
with an added comonomer (M), *e.g.*

$$\text{wwwF www} \quad +x\text{M} \quad \xrightarrow{\text{R}\cdot} \quad \text{wwF www} \quad \longrightarrow \quad F - M_x - M_x - F$$
$$\underset{(M)_x}{|}$$

$$\underset{F}{\text{www}} \quad +x\text{M} \quad \xrightarrow{\text{R}\cdot} \quad \underset{F(M)_x}{\text{www}} \quad \longrightarrow \quad F(M)_x - (M)_x F$$

Generation of the initiating radicals (R) arises from a
photochemical process of the types already listed in the
introduction to this survey. When the copolymerizing monomer
(M) produces a growing radical which terminates by combination, as indicated, crosslinking results. Alternatively, if
chain growth is limited by disproportionation or transfer
reactions, the product is a graft copolymer. Functionality
(F) of the preformed polymer will consist either of a copoly-
merizable unsaturated linkage type as, for example, in an
unsaturated polyester, or the locus of initiation for the
graft reaction.

(ii) Photoinduced chemical bonding between preformed
polymer molecules. Here the requirements are for the presence

in the polymer of either a group which *via* its photoexcited states, or complexes, can react with ground-state
molecules in other polymer molecules, or of a group which
will undergo photochemical decomposition (*e.g.* azido group)
so as to produce a highly reactive chemical intermediate.
Cyclodimerization of olefins is a particularly useful
example of such processes and sensitized dimerizations of,
for example, polymer-based cinnamic acid derivatives have
found wide application (45). The reactions may be represented in outline as

$$\sim\!\!A = B\!\!\sim \xrightarrow{\;h\nu\;} \sim\!\!A = B^{*}\!\!\sim$$

$$\xrightarrow{\sim\!\!A = B\!\!\sim\;\;} \begin{array}{c} \sim\!\!A - B\!\!\sim \\ |\quad\;\; | \\ \sim\!\!A - B\!\!\sim \end{array}$$

$$\begin{array}{c} | \\ A = B \end{array} \xrightarrow{\;h\nu\;} \begin{array}{c} | \\ A = B^{*} \end{array} \xrightarrow{\quad \begin{array}{c} A = B \end{array} \quad} \begin{array}{c} A - B \\ |\quad | \\ A - B \end{array}$$

Photoinduced electron-transfer reactions *via* encounter
complexes, exciplexes or charge-transfer complexes, afford
new ways of achieving the main polymer crosslinking reactions
(i) and (ii) above, with the particular advantages of improved
light absorption characteristics and the wide variety of
substrate types necessary to permit design of photosystems
fully compatible with, and stable in, the appropriate poly-
meric matrix.

The ketone-amine systems discussed in this survey represent just one class of photoactive combinations and, while
the ultimate objectives of such studies are the generation of
chemically reactive intermediates, it is clear that a more
detailed knowledge of photophysical phenomena will be
required for proper evaluation and screening of useful
systems.

ACKNOWLEDGEMENT

The experimental results presented in this survey were obtained during the graduate studies of R.A. Crellin, D.H. Iles, M.C. Lambert and M.D. Purbrick and it is a pleasure to acknowledge their initiative, interest and enthusiasm.

## REFERENCES

1. R.S. Mulliken, *J. Amer. Chem. Soc.* <u>74</u> , 811 (1952); *J. Phys. Chem.* <u>56</u>, 801 (1952).

2. E.M. Kosower, *Prog. Phys. Org. Chem.* <u>3</u>, 81 (1965).

3. A. Weller, *Pure Appl. Chem.* <u>16</u>, 115 (1968), and references therein.

4. M. Ottolenghi, *Accounts Chem. Res.* <u>6</u>, 153 (1973).

5. J. Hutchison and A. Ledwith, *Adv. Polymer Sci.* <u>14</u>, 49 (1974).

6. A. Ledwith *in* "Polymer Science" (E.B.H. Bawn, ed.), Vol. 2, MTP, London 1974.

7. H.G. Heine, H.J. Rosenkranz and H. Rudolph, *Angew. Chem. Int. Ed.* <u>11</u>, 974 (1972).

8. H. Block, A. Ledwith and A.R. Taylor, *Polymer (London)* <u>12</u>, 271 (1971).

9. J. Hutchison, M.C. Lambert and A. Ledwith, *Polymer (London)* <u>14</u>, 250 (1973).

10. S.G. Cohen, A. Parola and G.H. Parsons, *Chem. Rev.* <u>73</u>, 141 (1973).

11. N.C. Yang and J. Libman, *J. Amer. Chem. Soc.* <u>95</u>, 5783 (1973); T.R. Evans, *ibid.* <u>93</u>, 2081 (1971); N. Orbach, R. Potashnik and M. Ottolenghi, *J. Phys. Chem.* <u>76</u>, 1133 (1972); B.K. Selinger and R.J. McDonald, *Aust. J. Chem.* <u>25</u>, 897 (1972).

12. P. Hyde and A. Ledwith *in* "Molecular Complexes." (R. Foster, ed.) Vol. II. Logos Press, 1974.

13. S.G. Cohen and R.J. Baumgarten, *J. Amer. Chem. Soc.* <u>87</u>, 2996 (1965); <u>89</u>, 3471 (1967).

14. K. Yoshihara and D.R. Kearns, *J. Chem. Phys.* <u>45</u>, 1991 (1966).

15. L.A. Singer, *Tetrahedron Letters* 923 (1969); R.A. Caldwell, *ibid.* 2121 (1969); R.S. Davidson and P.F. Lambeth, *Chem. Commun.* 1098 (1969).

16. J.B. Guttenplan and S.G. Cohen, *Tetrahedron Letters* 2125 (1969).

17. G.A. Davis, P.A. Carapellucci, K. Szoc and J.D. Gresser, *J. Amer. Chem. Soc.* _91_, 2264 (1969).

18. S.G. Cohen and G. Parsons, *J. Amer. Chem. Soc.* _92_, 7603 (1970).

19. R.A. Caldwell and R.P. Gajewski, *J. Amer. Chem. Soc.* _93_, 532 (1971).

20. R.S. Davidson, P.F. Lambeth, J.F. McKellar, P.H. Turner and R. Wilson, *Chem. Commun.* 732 (1969); R.S. Davidson and R. Wilson, *J. Chem. Soc.* _B_ 71 (1970).

21. A. Ledwith and M.D. Purbrick, *Polymer (London)* _14_, 521 (1973).

22. R.S. Davidson and P.R. Steiner, *J. Chem. Soc., Perkin II* 1357 (1972).

23. C.H. Bamford and S. Brumby, *Makromol. Chem.* _105_, 122 (1967).

24. D. Valentine and G.S. Hammond, *J. Amer. Chem. Soc.* _94_, 3449 (1972).

25. R.S. Davidson and S.P. Orton, *Chem. Commun.* 209 (1974).

26. D. Braun and K.H. Becker, *Makromol. Chem.* _147_, 91 (1971).

27. M.R. Sandner, C.L. Osborne and D.J. Trecker, *J. Polymer Sci., A-1* _10_, 3173 (1972).

28. V.D. McGinniss and D.M. Dusek, Polymer Preprints, _15_, 480 (1974).

29. D.H. Davies, D.C. Phillips and J.D.B. Smith, *J. Org. Chem.* _14_, 2562 (1973); J.W. Breitenbach, O.F. Olaj and F. Wehrmann, *Monatsh. Chem.* _95_, 1007 (1964).

30. C.E.H. Bawn, A. Ledwith and Y. Shih-Lin, *Chem. Ind.*
    *(London)* 769 (1965); F.A. Bell, R.A. Crellin, H. Fujii
    and A. Ledwith, *Chem. Commun.* 251 (1969).

31. R.A. Carruthers, R.A. Crellin and A. Ledwith, *Chem.*
    *Commun.* 252 (1969).

32. S. Arai, A. Kira, M. Imamura, Y. Tabata and K. Oshima,
    *J. Polymer Sci., Polymer Letters* Ed. *10*, 295 (1972).

33. A. Ledwith, *Accounts Chem. Res.* *5*, 133 (1972).

34. R.A. Crellin, M.C. Lambert and A. Ledwith, *Chem. Commun.*
    682 (1970).

35. K. Tada, Y. Shirota and H. Mikawa, *J. Polymer Sci.,*
    *Polymer Chem. Ed.* *11*, 2961 (1973).

36. K. Tada, Y. Shirota and H. Mikawa, *Macromolecules* *6*,
    9 (1973).

37. Y. Taniguchi, Y. Nishina and N. Mataga, *Bull. Chem.*
    *Soc. Japan* *46*, 1646 (1973); Y. Shirota, K. Kawai,
    N. Yamamoto, K. Tada, T. Shida, H. Mikawa and H. Tsubomu
    *Bull. Chem. Soc. Japan* *45*, 2683 (1973); M. Yamamoto,
    M. Ohoka, K. Kitagawa, S. Nishimoto and Y. Nishijima,
    *Chem. Letters (Japan)* 745 (1973).

38. S. Farid and S.E. Shealer, *Chem. Commun.* 677 (1973).

39. T.R. Evans, R.W. Wake and O. Jaenicke *in* "The Exciplex."
    (M. Gordon, W.R. Ware, P. de Mayo and D.R. Arnold, eds.)
    Academic Press, New York, 1975.

40. J.M. Fritsch, T.P. Layloff and R.N. Adams, *J. Amer. Chem*
    *Soc.* *87*, 1724 (1965).

41. J.F. Ambrose and R. F. Nelson, *J. Electrochem. Soc.*
    *115*, 1159 (1968).

42. C.K. Mann and K.K. Barnes, "Electrochemical Reactions
    in Nonaqueous Systems." Marcel Dekker, New York, 1970.

43. D.H. Iles, Ph.D. Thesis, University of Liverpool (1969).

44. M.C. Lambert, Ph.D. Thesis, University of Liverpool (1974).

45. J.L.R. Williams, *Fortschritt. der chemischen Forschung (Topics in Current Chemistry)* _13_, 227 (1969); G.A. Delzenne, *J. Makromol. Sci., Revs. Polymer Technol. D1(2)*, 185 (1971).

SPECTROSCOPIC STUDIES OF NONLUMINESCENT EXCIPLEXES.

ELECTRON TRANSFER AND EXCIPLEX FORMATION FROM

TRIPLET STATES OF ANTHRACENE AND METALLOPORPHYRINS

DAVID G. WHITTEN, J.K. ROY AND FELIX A. CARROLL

*Department of Chemistry, University of North Carolina*

*Chapel Hill, North Carolina   27514*

INTRODUCTION

Photochemical reactions of metalloporphyrins and related compounds have been the subject of extensive investigation largely because of the importance of these compounds in photobiological and electron-transport processes. Special emphasis has been placed on photoredox phenomena of porphyrins due to the role of chlorophyll in the energy trapping process of photosynthesis. Relatively early, Livingstone and co-workers found that fluorescent singlets of chlorophyll and various metalloporphyrins were quenched by electron-deficient compounds such as quinone, aryl nitro compounds and nitroso compounds (1 - 3). Subsequently, it has been found that excited porphyrins can serve as both formal electron donors and electron acceptors in a variety of reactions which include both electron and atom transfer (4 - 8).

Our own work in this area (9 - 13) has dealt largely with the reaction of excited metalloporphyrins with electron-deficient compounds such as quinones and nitroaromatics. By using reactive nitroaromatics such as *p*-nitrostilbene and 4,4'-dinitrostilbene as quenchers for metalloporphyrins, we were able to probe various decay routes in the complexes formed from excited singlets and triplets of the metallo-porphyrins (9, 10). More recently our work has included spectroscopic studies of the exciplexes formed from triplet states of metalloporphyrins with nitroaromatics (11 - 13). In the present paper we report an extension of this work which includes a greater variety of substrates and quenchers. Results of this study suggest that exciplex formation from excited triplets is a fairly general phenomenon. Our investigation of factors governing formation and decay of these triplet exciplexes has enabled some conclusions to be drawn about their structure and binding forces; a significant aspect of this study is the finding that, for these exciplexes, charge-transfer interactions, although important, must play a limited role.

EXPERIMENTAL

The preparation and purification of substrates and quenchers used in this study have been previously reported (13). The methods and apparatus for the flash-spectroscopic investigation have also been previously described (10, 14).

RESULTS

We have previously reported the quenching of excited singlet and triplet states of zinc etioporphyrin I (Zn etio) by nitroaromatics in nonpolar solvents (9, 10). In extensions

f this work, we find that aliphatic chloro compounds such
.s hexachloroethane (HCE) and carbon tetrachloride as well
s chloroaromatics such as DDT [1,1,1-trichloro-2,2-bis-
p-chlorophenyl)ethane] and p-dichlorobenzene also serve as
uenchers for Zn etio, other porphyrins and metalloporphyrins,
nd anthracene (15, 16). Table 1 lists data for the quench-
ng of Zn etio fluorescence by several of these quenchers

TABLE 1

*ate Constants for Quenching of Zinc Etioporphyrin Excited*
*inglets in Benzene*

| Quencher | $k_q \times 10^{-8}$ $M^{-1}$ $sec^{-1}$[a] | Quencher Reduction Potential[b] (volts vs SCE) |
|---|---|---|
| lexachloroethane | 9 | 0.62 |
| DT | 4 | 0.88 |
| -Dichlorobenzene | 2 | 2.05 |
| -Benzoquinone | 140 | 0.97 |
| is-4-Nitrostilbene | 72 | |
| litrobenzene | 38 | 1.14 |
| -Nitroaniline | 49 | 1.38 |
| NT | 33 | 1.22 |
| rans-4-Nitrostilbene | 90 | 1.11 |

[a] Values at 25° [b] Measured in acetonitrile or dichloromethane.
Je thank Mr. G. Brown for these measurements.

while table 2 gives extensive data for the quenching of a
variety of substrate triplets. Rate constants listed in
:able 1 were obtained from the usual Stern-Volmer
relationship while the triplet quenching data were obtained
'rom lifetimes measured flash spectroscopically.

# TABLE 2

*Experimental and Calculated Values for Triplet Quenching and Transient Behavior in Benzene*

| Substrate Quencher System | $k_t \times 10^{-8}$ $M^{-1}$ sec$^{-1}$ [a] | Substrate Oxidation Potential [b] $-E_{1/2}$ (volts vs SCE) | Quencher Reduction Potential [b] $-E_{1/2}$ (volts vs SCE) | Calculated Energy of $S \cdot \bar{Q}^{-}$ State (kcal/mole) [c] | Transient Detected [d] |
|---|---|---|---|---|---|
| Zn Etio I – 9-nitroanthracene | 45 | 0.63 | – | – | – |
| Zn Etio I – p-dinitrobenzene | 42 | 0.63 | 0.69[e] | 32.7 | – |
| Zn Etio I – p-nitrobenzaldehyde | 34 | 0.63 | 0.89 | 37.4 | – |
| Zn Etio I – cis-4,4'-dinitrostilbene | 31 | 0.63 | 1.00 | 39.8 | SQ₂ |
| Zn Etio I – PNT | 30 | 0.63 | 1.22 | 44.9 | SQ₂ |
| Zn Etio I – p-chloronitrobenzene | 24 | 0.63 | 1.16 | 43.4 | – |
| Zn Etio I – trans-1-(4-nitrophenyl)-2-phenylpropene | 20 | 0.63 | – | – | – |
| Zn Etio I – nitrobenzene | 20 | 0.63 | 1.14 | 43.2 | SQ |
| Zn Etio I – trans-4-nitrostilbene | 16 | 0.63 | 1.11 | 42.4 | SQ₂ |
| Zn Etio I – α-bromo-PNT | 15 | – | – | – | – |
| Zn Etio I – p-nitroanisole | 10 | 0.63 | 1.26 | 45.9 | SQ₂ |
| Zn Etio I – p-nitroaniline | 1.2 | 0.63 | 1.38[f] | 48.7 | – |
| Zn Etio I – hexachloroethane | 1.4 | 0.63 | 0.62[f] | 31.1 | SQ |

| | | | | |
|---|---|---|---|---|
| Zn Etio I – DDT | 0.82 | 0.63 | 0.88 | 37.1 | SQ |
| Zn Etio I – hexachlorobenzene | 0.67 | 0.63 | 1.44[h] | 50 | SQ2 |
| Zn Etio I – p-dichlorobenzene | 0.4 | 0.63 | 2.05[h] | 64 | SQ2 |
| Zn Etio I – carbon tetrachloride | 0.3 | 0.63[g] | >2.20 | >68 | – |
| Zn OEP – PNT | 19 | 0.53[g] | 1.22 | 42.7 | – |
| Zn OEP – p-chloronitrobenzene | 13 | 0.53 | 1.16 | 41 | – |
| Zn OEP – p-nitroanisole | 8 | 0.53 | 1.26 | 43.6 | – |
| Zn OEP – p-nitroaniline | 0.6 | 0.53 | 1.38 | 46.3 | – |
| Zn OEP – hexachloroethane | 7.1 | 0.53 | 0.62 | 28.8 | SQ |
| Zn OEP – DDT | 0.3 | 0.53 | 0.88 | 34.8 | SQ |
| Zn OEP – hexachlorobenzene | 0.4 | 0.53 | 1.44 | 47.7 | – |
| Mg Etio I – nitrobenzene | 18 | 0.46 | 1.14 | 39.2 | – |
| Mg Etio I – PNT | 14 | 0.46 | 1.22 | 41 | SQ2 |
| Mg Etio I – p-nitroanisole | 7 | 0.46 | 1.26 | 42 | – |
| Mg Etio I – cis-4,4'-dinitrostilbene | – | 0.46 | 1.00 | 36.0 | SQ |
| Etio I – PNT | 0.035 | 0.81 | 1.22 | 49.8 | SQ2 |
| Etio I – nitrobenzene | 0.0014 | 0.81 | 1.14 | 48 | SQ2 |
| Etio I – p-nitroanisole | 0.35 | 0.81[i] | 1.26 | 50.7 | SQ2 |
| Anthracene – PNT | 0.004 | 1.09[i] | 1.22 | 55.6 | SQ2 |
| Anthracene – hexachloroethane | 0.14 | 1.09 | 0.62 | 41.7 | SQ2 |
| Anthracene – DDT | 0.10 | 1.09 | 0.88 | 47.7 | SQ2 |
| Anthracene – hexachlorobenzene | 0.05 | 1.09 | 1.44 | 60.5 | – |
| Anthracene – p-dichlorobenzene | 0.03 | 1.09 | 2.05 | 74.7 | – |
| Anthracene – carbon tetrachloride | 0.16 | 1.09 | >2.20 | 78.1 | – |

[a] Values at 25°. [b] Measured in acetonitrile or dichloromethane. We thank Mr. G. Brown for these measurements. [c] From equation [4], see ref. 26. [d] Determined from kinetic behavior at high [quencher], see text. [e] A. H. Maki and D.H. Geske, J. Chem. Phys. 33, 825 (1960). [f] M. von Stackelberg and W. Stracke, Z. Electrochem. 53, 118 (1949). [g] J.-H. Fuhrhop and D. Mauzerall, J. Amer. Chem. Soc. 91, 4174 (1969). [h] J. W. Sease, F.G. Burton and S.L. Nikol, J. Amer. Chem. Soc. 90, 2595 (1968). [i] E.S. Pysh and N.C. Yang, J. Amer. Chem. Soc. 85, 2124 (1963).

The triplet quenching process, which has been investigated much more extensively than the singlet quenching, was
also studied as a function of temperature over the range
$20 - 78^{\circ}$ C. Reasonable Arrhenius correlations were obtained;
activation energies obtained from the slopes of these plots
and $\Delta S$ values calculated from the experimental preexponential
factors are listed in table 3. [The $k_q^t$ values were corrected
for changes in the triplet lifetimes, which were significant
over this temperature range, but not for changes (nearly
negligible) in solvent viscosity.] Activation energies are

TABLE 3

*Activation Parameters for Quenching of Triplet Zinc*

*Etioporphyrin and Anthracene by Nitroaromatics and Chloro*

*Compounds*

| Substrate-Quencher System | $E_a$ (kcal/mole)[a] | $-\Delta S$ (eu) |
|---|---|---|
| Zn Etio I - p-nitrobenzaldehyde | 0.35 | 17 |
| Zn Etio I - PNT | 0.48 | 17 |
| Zn Etio I - *p*-chloronitrobenzene | 0.46 | 17 |
| Zn Etio I - nitrobenzene | 0.94 | 16 |
| Zn Etio I - *trans*-4-nitrostilbene | 1.33 | 17 |
| Zn Etio I - *p*-nitroanisole | 0.5 | 14 |
| Zn Etio I - *p*-nitroaniline | 0.73 | 17 |
| Zn Etio I - hexachloroethane | 1.2 | 24 |
| Zn Etio I - DDT | 1.4 | 23 |
| Zn Etio I - hexachlorobenzene | 1.5 | 21 |
| Zn Etio I - *p*-dichlorobenzene | 3.2 | 16 |
| Zn Etio I - carbon tetrachloride | 3.3 | 13 |
| Anthracene - PNT | 4.7 | 19 |
| Anthracene - hexachloroethane | 1.3 | 24 |
| Anthracene - DDT | 3.0 | 24 |
| Anthracene - hexachlorobenzene | 1.4 | 26 |
| Anthracene - *p*-dichlorobenzene | 2.6 | 22 |

[a] Calculated from temperature dependence of $k_q^t$, corrected for
change in substrate triplet lifetime with temperature.

enerally smaller for nitroaromatics than for the chloro
ompounds and somewhat higher for anthracene than for Zn etio.
ntropies of activation are all negative and in the range
14 to -26 eu.

Quenching of substrate excited singlet states occurs at
igh quencher concentration while triplet quenching is gen-
rally complete at much lower quencher concentration levels.
hus it is relatively easy to investigate, selectively, by
lash spectroscopy the transients formed in each process. In
enzene it was found that for the metalloporphyrins as sub-
trates, high quencher concentrations (where significant
luorescence quenching occurs) led to no detectable transients
aving lifetimes long enough to measure with microsecond
lash-spectroscopic techniques. With lower quencher con-
entrations, the transient signals due to triplet metallo-
orphyrin or anthracene (at 420 nm and 430 nm respectively)
howed first a reduction in lifetime with increase in
uencher concentration. However, at moderate quencher con-
entration, where triplet quenching should be complete but
inglet quenching should be negligible, new transients are
bserved. In most cases these transients have absorption
pectra similar, but not identical, to those of the substrate
riplets. Figures 1 and 2 compare spectra obtained for
ransients with no quencher and with moderate quencher con-
entration for typical combinations. Figures 3 and 4 show
he two patterns observed for transient lifetime as a function
f quencher concentration. For Zn etio with HCE (figure 3),
ow concentrations of HCE quenched the porphyrin triplet
ith a rate constant $k_q^t = 1.4 \times 10^8$ $M^{-1}$ sec$^{-1}$. At higher HCE
oncentrations, the transient lifetime reached a limiting
alue of *ca.* 40 μsec. The transient observed at these higher
oncentrations had a spectrum (figure 1) only slightly

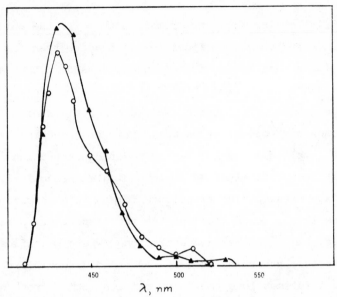

λ, nm

*Fig. 1   Transient absorption spectra for zinc etio-porphyrin I in benzene (○) and for zinc etioporphyrin I - HCE [0.01 M] in benzene (▲).*

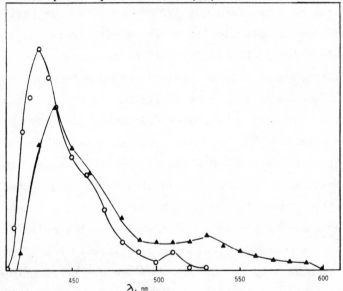

λ, nm

*Fig. 2   Transient absorption spectra for zinc etio-porphyrin I in benzene (○) and for zinc etioporphyrin I - PNT [0.01 M] in benzene (▲).*

254

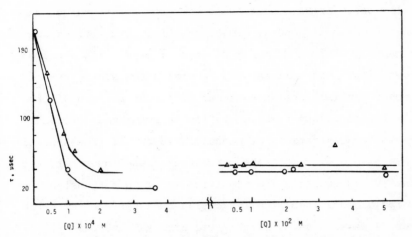

*Fig. 3 Plot of transient lifetime vs quencher con-
centration for zinc etioporphyrin I - HCE (O) and for
zinc etioporphyrin I - DDT. (Δ).*

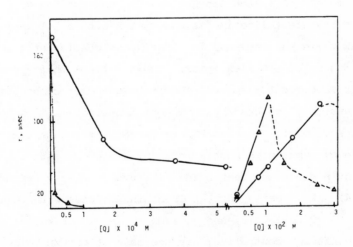

*Fig. 4 Plot of transient lifetime vs quencher con-
centration for zinc etioporphyrin I - PNT (Δ) and for
zinc etioporphyrin I - p-dichlorobenzene (O).*

different from that of the unquenched Zn etio triplet.
Similar behavior was observed for several substrate-quencher
combinations;  in general limiting transient lifetimes are in
the range 20 - 70 μsec.  For Zn etio triplets with *p*-nitro-
toluene (PNT), as well as with several other substrate-
quencher combinations, a somewhat different pattern of transient lifetime-quencher concentration behavior was observed.
Here low concentrations of quencher reduce the transient lifetime;  however, at moderate concentrations of quencher, a new
transient is detected and the lifetime of this transient
absorption increases with increasing quencher concentration.
Figure 4 shows results for Zn etio with PNT and *p*-dichloro-
benzene as quenchers.  Although more extensive data on
transient lifetimes at high quencher concentration are available for the metalloporphyrins, similar behavior is observed
for triplet anthracene with both groups of quenchers.  As
will be developed in the discussion, the observation of a
constant transient lifetime is associated with a 1:1 (SQ)
exciplex while the increase in lifetime is consistent kinetically with a 1:2 (SQ$_2$) exciplex.  Table 2 lists the type of
exciplex observed for several substrate-quencher combinations.

Transients produced by interaction of triplet Zn etio
with nitroaromatics such as PNT in benzene are themselves
quenched by substances such as oxygen, azulene, tetracene and
perylene in processes that evidently involve energy transfer
(table 4) (12).  Rates for the quenching process with tetracene (E$_t$ = 29 kcal/mole) and azulene (E$_t$ = 31 kcal/mole) are
nearly diffusion controlled;  in the case of the Zn etio-PNT
exciplex, the tetracene triplet could be detected flash
spectroscopically as a product of the quenching process (12).

The quenching process and transient behavior for several
substrates and quenchers have also been investigated in more

## TABLE 4

*Quenching of the Transient from Zinc Etioporphyrin and*

*p-Nitrotoluene[a]*

| Quencher | Benzene | $k_q, M^{-1} \sec^{-1}$ | N-Methylformamide |
|----------|---------|-------------------------|-------------------|
| Azulene | $2.6 \times 10^9$ | | no quenching[b] |
| Tetracene | $5.1 \times 10^8$ | | no quenching |
| Perylene | $1.3 \times 10^9$ | | |

[a] Degassed solutions, $k_q$ values determined by flash spectroscopy. Transient generated from $5 \times 10^{-5}$ M zinc etioporphyrin I and 0.01 M PNT. [b] Transient half-life not reduced.

polar solutions. Results obtained in the more polar solutions indicate some rather surprising differences in transient behavior, even though the excited-state quenching patterns are not dramatically changed.

Rate constants for fluorescence quenching of Zn etio and Mg etio by nitroaromatics in different solvents have been previously reported (9). These increase slightly as the reaction is studied in more polar solvents. For example, with Zn etio and nitrobenzene there is a two-fold increase in $k_q^s$ as the reaction solvent is changed from benzene to ethanol and, for Mg etio with nitrobenzene, $k_q^s$ increases about 10% as the solvent is changed from benzene to acetonitrile. The effect of solvent on triplet quenching rates has not been extensively investigated; however, for Zn etio with PNT as quencher, $k_q^t$ values are $3 \times 10^9$, $2 \times 10^8$, and $1.3 \times 10^8$ $M^{-1}$ $\sec^{-1}$ for the solvent series benzene, 0.01 M N-methylformamide in benzene and pure N-methylformamide. The effect of solvent on the quenching of anthracene excited states has not yet been investigated. For the metalloporphyrins it is difficult to identify trends or draw conclusions regarding the solvent effect since the data are limited and specific solvation or complexation at the metal, perhaps, plays some role in the quenching process. Nevertheless, quenching of both excited singlets and triplets does occur in the more polar solvents and the quenching rates are generally within an order of magnitude of those in benzene or aliphatic hydrocarbons.

The effect of change in solvent on transient lifetime has been investigated for several substrate-quencher combinations. In most cases the following pattern is observed: addition of a polar solvent causes first a sharp decrease in the lifetime of the transients generated from substrate triplets followed by a rapid increase in decay time upon subsequent addition of the polar solvent. The increase in decay time is accompanied by a change in kinetic behavior from first- to second-order decay as well as by changes in the transient absorption spectrum. The long-lived transients produced in the more polar media are not quenchable by oxygen or tetracene and, in general, it is found that in more polar media identical transients can be generated by the quenching of excited singlet or triplet states of the substrate.

Figure 5 shows the variation in $t_{1/2}$ for transient decay as a function of added polar solvent for the transients generated by interaction of PNT and nitrobenzene with Zn etio triplet. Similar behavior was observed for Mg etio and free-base etioporphyrin with several nitroaromatics. In these cases, the absorption spectrum of the transient generated in mixtures of high solvent polarity showed absorption both in t

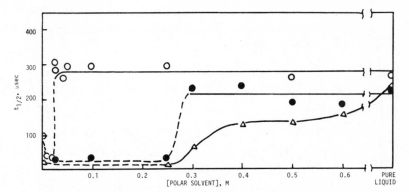

Fig. 5 *Transient half-life for zinc etioporphyrin
complexes as a function of added polar cosolvent to
benzene solutions.* [○] *Zinc etio - PNT (0.01 M)
with added N-methylformamide;* [●] *Zinc etio -
PNT (0.01 M) with added ethanol;* [▲] *Zinc etio -
nitrobenzene (0.01 M) with added ethanol.*
(——) *Region where second order decay observed.*
(---) *Region where first order transient decay
observed.*

420 - 550 nm region (where the exciplexes generated in benzene
absorb) and in the 590 - 660 nm region.  These spectra
(figure 6) resemble those of π-cation radicals generated by
one-electron oxidation of zinc and magnesium porphyrins
(17, 18).  Since Zn etio and nitroaromatics produce transients
in ethanol having nearly identical half-lives and spectra as
those produced from Zn etio and p-benzoquinone, a system for
which ESR spectroscopy has shown that electron transfer to yield
ions is the dominant reaction (19, 20), we infer that the transients produced in the polar solvents and solvent mixtures are
the metalloporphyrin radical cation and the nitroaromatic
radical anion.  If free ions are the product of the quenching
process, it would be expected that these ions should return to
the ground states of the starting materials in a (second-order)
diffusion-controlled recombination.  A half-life of

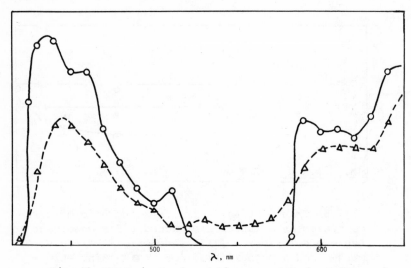

*Fig. 6 Transient absorption spectra in polar sol-
vents. [O] Zinc etio - nitrobenzene in 0.5 M
ethanol in benzene. [△] Zinc etio - PNT in N-methylformamide.*

*ca.* 200 µsec is predicted from an estimation that the initial
concentration of ions is $10^{-6}$ M and $k_{diff}$ = 5 × $10^9$ $M^{-1}$ $sec^{-1}$.
This is in reasonable agreement with the values of 180 - 300
µsec obtained for a variety of different systems using triplet Zn etio as the excited donor. In agreement with the
anticipated slower diffusion rates for ions in more polar
solvents, $t_{\frac{1}{2}}$ values for Mg etio and for Zn etio with nitroaromatics are found to be longer in N-methylformamide (300 -
400 µsec) than in ethanol (100 - 200 µsec) even though the
initial transient absorption intensities are about the same.

For the reaction of Zn etio triplets with chloro compounds in polar solvents, ion formation is evidently not
completely efficient. Various patterns are observed which
depend strongly upon the specific quencher and solvent system.
In several cases (HCE, DDT and hexachlorobenzene), there is

some production of a long-lived transient absorbing in the range of 590 - 660 nm and exhibiting second-order decay. However, absorption intensities in this range are generally weaker and the half-lives longer (*e.g.* 730 μsec for HCE in N,N-dimethylformamide) indicating that ion formation in these systems is somewhat inefficient. In several of these systems, a transient absorbing at shorter wavelengths (420 - 500 nm) and decaying by a first-order process is also observable. Transients formed from anthracene with nitroaromatics and chloro compounds in polar solvents have not yet been investigated. However, it has been found that other aromatic hydrocarbons such as perylene react in the excited state with nitroaromatics such as PNT in acetonitrile and N-methylformamide to yield transients exhibiting second-order decay with absorption spectra similar to that of the perylene cation ($\lambda_{max}$ 540 nm).

## DISCUSSION

For the substrate-quencher systems under investigation in this study, it is clear that rather different phenomena are observed in polar and in nonpolar solutions. In polar solvents or solvent mixtures, it appears that net electron transfer to give substrate radical cations and quencher radical anions is the end result of the quenching process in several cases. Such a result is not surprising in view of the fact that most of the quenchers are fairly good oxidants; indeed, electron transfer from excited triplets of other substrates to electron acceptors has been observed in polar media in several cases (21 - 22). The observation that the transients produced in polar solvent systems follow second-order decay kinetics and are not quenchable by oxygen or

substances having low lying triplet states suggests that
they are free ions rather than exciplexes or other electronically excited species. In contrast, the transients produced
in nonpolar solvents are evidently true exciplexes—electronically excited complexes. However, they do not appear to be
adequately described by most previously developed models
(23 - 26).

To develop a better description of the exciplexes formed
between substrate triplets and the quenchers used in this
study, it is useful to consider first the quenching process
and then the transients themselves and their behavior. The
model developed by Weller and coworkers (23, 24, 27, 28)
suggests that exciplexes are charge-transfer complexes
(radical ion pairs) which may decay to free ions or to ground
states. For such complexes, it has been suggested that $k_q$
should vary linearly with quencher reduction potential if it
is presumed that the excited state is the electron donor and
the quencher is the electron acceptor in forming the complex
(equation [1]).

$$S* + Q \xrightarrow{\phantom{xx}k_q\phantom{xx}} (S^{\dot{+}}...Q^{\dot{-}}) \qquad [1]$$

For the quenching of Zn etio singlets (table 1), it is clear
that there is not a good correlation of this type. For the
nitroaromatics (which quench at rates close to diffusion-
controlled), there appears to be little correlation of reduction potential with $k_q^S$ ; the three chloro compounds studied
quench in the sequence anticipated from the reduction
potential but the difference in quenching rates is small compared to the differences in quencher reduction potential.
Moreover there is no correlation whatsoever when the nitroaromatics and chloro compounds are taken together as a series
of singlet quenchers.

For the quenching of substrate triplets in benzene there is a somewhat better correlation of $k_q$ values with quencher reduction potential. For Zn etio with nitroaromatics (figure 7), and with chloro compounds (figure 8), plots of $\log k_q^t$ vs $E_{\frac{1}{2}}$ for quencher reduction show good linearity. However, distinctly different correlations are observed between the two classes of quenchers. This different pattern of reactivity emerged for all the systems examined. In figure 9, $\log k_q^t$ vs quencher $E_{\frac{1}{2}}$ is plotted for anthracene with chloro compounds. Although the overall rate constants are much lower in these cases, the slope is very close in value to that obtained for Zn etio with the same quenchers in figure 8. The results observed here are somewhat reminiscent of results obtained by Carroll, McCall and Hammond (29) for the quenching of 1,4-dimethoxybenzene fluorescence by various groups of quenchers. We infer the same conclusion that these authors made, namely that charge-transfer or electron donor-acceptor interactions play a key role in the quenching process but $k_q$ must depend on other factors in addition to simple electron-transfer parameters.

Weller and coworkers (23, 24, 27, 28) have found good correlations between the energies of "pure" ($S^+$, $Q^-$) states (CT states) and measured polarographic (in acetonitrile) oxidation and reduction potentials in several solvents. The energy of this state may be estimated in benzene to be given by equation [2] (24, 30, 31). Calculated energies of these

$$E_{(S^+, Q^-, benzene)} = E_{\frac{1}{2}(S/S^+)} - E_{\frac{1}{2}(Q^-/Q)} + 0.10$$

$$\pm \ 0.10 \ eV \tag{2}$$

states for the substrate-quencher combinations investigated in this work are listed in table 2. Triplet energies of the

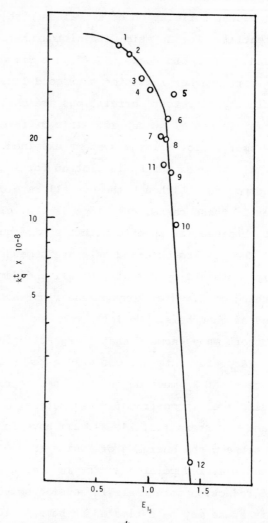

Fig. 7 Plot of log $k_q^t$ vs quencher reduction potential (volts vs S.C.E.) for zinc etioporphyrin and nitroaromatics. Numbers refer to the following quenchers: 1, 9-nitroanthracene; 2, p-dinitrobenzene; 3, p-nitrobenzaldehyde; 4, cis-4,4'-dinitrostilbene; 5, PNT; 6, p-chloronitrobenzene; 7, trans-1-phenyl-2-(p-nitrophenyl)propene; ·8, nitrobenzene; 9, α-bromo-p-nitrotoluene; 10, p-nitroanisole; 11, trans-4-nitrostilbene; 12, p-nitroaniline.

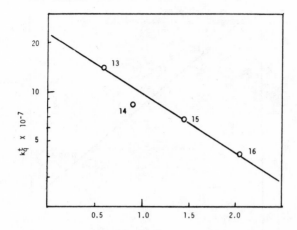

Fig. 8  Plot of log $k_q^t$ vs quencher reduction potential (volts vs S.C.E.) for zinc etioporphyrin and
chloro compounds. Numbers refer to the following
quenchers: 13, HCE; 14, DDT; 15, hexachlorobenzene;
16, p-dichlorobenzene.

substrates used in this study are as follows:  Zn etio,
41 kcal/mole (9, 14);  Mg etio, 41 kcal/mole and anthracene,
42 kcal/mole (32).  From the energies listed in table 2, it
can be seen that, in several cases, the "pure" charge-
transfer state lies several kcal/mole above the substrate
triplet level.  For example, for Zn etio-PNT, the calculated
CT state energy is 44.9 kcal/mole;  an estimated $k_q^t$ based on
a minimum 3.8 kcal/mole activation energy would be  ca.
$10^6$ - $10^7$ $M^{-1}$ $sec^{-1}$, yet the actual value is higher by between
two and three orders of magnitude.  Similarly for anthracene-
PNT, the energy of the CT state is estimated to be 55.6 kcal/
mole which is 13.6 kcal/mole above that of the anthracene
triplet.  A minimum activation energy this large should
lower $k_q^t$ nearly ten orders of magnitude below diffusion-
controlled and yet the actual value is $4.4 \times 10^5$ $M^{-1}$ $sec^{-1}$.

265

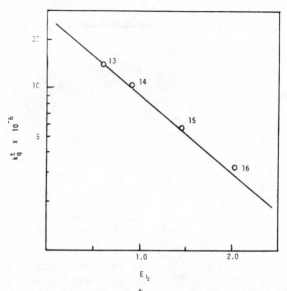

Fig. 9 Plot of log $k_q^t$ vs quencher reduction potential (volts vs S.C.E.) for anthracene and organic
chloro compounds. Numbers refer to the compounds
listed under figure 8.

Clearly, these simple considerations indicate that, in cases
where the calculated CT energy is above that of the substrate
triplet, it is unlikely that the exciplex formed can be a
pure CT state. Moreover, evaluation of the activation parameters in table 3 indicates that in several additional cases
the exciplexes must not be "pure" CT states. The usual model
for formation and decay of simple 1:1 exciplexes is given by
equations [3] - [5], as

$$S* + Q \underset{k_{-1}}{\overset{k_1}{\rightleftharpoons}} (SQ)* \qquad [3]$$

$$SQ* \overset{k_2}{\longrightarrow} S + Q \qquad [4]$$

$$k_q(obs) = \frac{k_1 k_2}{k_{-1} + k_2} \qquad [5]$$

Limiting situations are given by equations [6] and [7].

$$\text{if } k_{-1} \gg k_2, \ k_{q(obs)} = \frac{k_1}{k_{-1}} \ k_2 = K_{eq} \ k_2 \qquad [6]$$

$$\text{if } k_2 \gg k_{-1}, \ k_{q(obs)} = \frac{k_1}{k_2} \ k_2 = k_1 \qquad [7]$$

For the substrate-quencher combinations investigated, activation parameters (table 3) are generally similar; $E_a$ is relatively small and positive while $\Delta S$ is large and negative. The $\Delta S$ values appear much more consistent with a mechanism involving rate-limiting formation of a fairly ordered complex (33) (equation [7]) than with the alternative situation given by equation [6]. For the substrate-quencher combination giving rise to simple 1:1 (SQ)* exciplexes, $k_2$ can be estimated, from the limiting lifetime at high quencher concentration, to be generally between $10^4 \ M^{-1} \ sec^{-1}$ and $5 \times 10^4 \ M^{-1} \ sec^{-1}$. If this value is compared with the observed $k_q^t$ values in equations [6] and [7], it becomes evident that, in either limiting situation, $k_{-1}$ must be much smaller than $k_1$ and hence $K_{eq}$ for exciplex formation is generally very large for the combinations investigated. This suggests that the true exciplex energy is substantially lower than the substrate triplet. Therefore, in all but a few cases (see below), the exciplexes formed between these substrates and quenchers are not pure CT states and, moreover, they must lie substantially lower in energy than pure CT states.

From the above analysis, a major question that develops is: if charge-transfer interactions play only a limited role in stabilizing these exciplexes, what other factors are important? It appears clear that exciton resonance involving dipole-dipole terms must play an insignificant role in the

quenching process. That is, the contribution of forms such
as S...Q* in equation [8] probably is very small, since

$$S* + Q \longrightarrow (SQ)* \longleftrightarrow (S*....Q) \longleftrightarrow (S....Q*) \qquad [8]$$

there is practically no correlation between $k_q$ values and
singlet and triplet energies of the quenchers (36). Considerable insight into the structure of these exciplexes and the
factors governing their formation can be obtained by
examination of the exciplex absorption spectra. Examination
of figures 1 and 2 shows that, while not identical, the
exciplex spectra resemble closely those of the unquenched
substrate triplets; they do not resemble the reported
spectra of $S^+$ or $Q^-$ ions (18). It can be concluded, therefore, that the energy levels of the substrate excited triplet responsible for the triplet → triplet transition have
not been greatly altered or are altered similarly in going
from S* to SQ* (38).

An attractive model for this type of exciplex formation
is one in which quencher molecules associate adiabatically
with the substrate triplet only through a loose overlap
of molecular orbitals with a small degree of S to Q charge
transfer. These exciplexes differ from pure CT states in
that complete electron transfer certainly does not occur
and the orbitals involved in the donor-acceptor interaction
*need* not be the singly populated orbitals resulting from
electronic excitation. This model could permit exciplex
binding energy (and complex-formation rates) to be correlated
with electron affinity data, but the fate of the exciplex
might be governed by entirely different factors. Moreover
this type of interaction would not necessarily alter the
substrate excited state greatly and thus spectral characteristics of the transient could be similar to those of the

substrate. An analogy for this model is the acid-base reactions of excited states such as protonation of the $\pi,\pi*$ singlet of acridine or phenoxide. Radiationless decay of the complex would probably be governed primarily by decay modes available to the substrate unless a polar environment allowed ion-pair formation or specific pathways (e.g. intersystem crossing induced by a heavy atom substituent) were introduced by the quencher.

It is perhaps interesting to consider this model with respect to the different types of exciplex observed for various substrate-quencher combinations in this study. As mentioned previously, two patterns of variation of complex lifetime with quencher concentration have been observed. A constant lifetime at high [quencher] is consistent with 1:1 exciplex having $\tau = 1/k_2$. The second pattern observed is a linear increase in $\tau$ with quencher concentration; we have previously shown this to be consistent with the formation of a 1:2 ($SQ_2*$) exciplex as shown in equations [4] and [9].

$$SQ* \xrightarrow{k_2} S + Q \qquad\qquad [4]$$

$$SQ* + Q \underset{k_{-3}}{\overset{k_3}{\rightleftharpoons}} SQ_2* \qquad\qquad [9]$$

If $k_2$ is the predominant decay channel for the exciplex, it can be shown that equation [10] applies.

$$\tau = \frac{1}{k_{obs}} = \frac{1}{k_{-3}} + \frac{k_3[Q]}{k_{-3}k_2} \qquad\qquad [10]$$

From the data in table 2, it may be seen that 1:1 exciplexes are observed (with the exception of Zn etio-nitrobenzene) where the calculated CT energy is substantially below the substrate triplet energy. The 1:2 exciplexes are generally the observed transients in the other cases examined thus far.

Although no evidence regarding the structure of 1:2 complexes is available, it appears reasonable that a planar aromatic hydrocarbon or metalloporphyrin complexed weakly on one side might easily complex on the other side to a second quencher molecule. It is conceivable that, in a complex involving only very weak donor-acceptor interactions, residual donor ability might persist in the substrate permitting this to occur. On the other hand, in a complex involving stronger electron donation (or more nearly pure CT complex formation), the substrate would have less residual electron-donating strength to complex to a second quencher. It appears reasonable that three-component exciplexes can best occur in cases where there are only weak donor-acceptor interactions. Whether the 1:1 complexes formed in this study are "pure" CT states is, as yet, unresolved, but it appears consistent that stronger donors and acceptors or a more polar environment should disfavor the three-component exciplexes.

The interactions between excited substrate and quenchers in nonpolar media where some sort of limited donor-acceptor interaction occurs, as exemplified by the present results, should probably be a fairly general phenomenon, especially for relatively long-lived triplet states. An interesting example, which may be related to the present phenomenon, is the gas-phase excited-state quenching reported by Thayer and Yardley (41) where quenching rate correlations with electron affinity are observed yet the quenching process almost surely does not involve electron transfer. Our results indicate that similar interactions can occur in nonpolar solvents in solution. It is easy to conceive of a continuum extending from specifically detectable quenching to subtle effects of solvent on excited-state lifetimes.

ACKNOWLEDGEMENTS

We are grateful to the U.S. Public Health Service (Grant No. GM 15238) and the University of North Carolina Materials Research Center for support of this work.

## REFERENCES

1. R. Livingston and C.L. Ke, *J. Amer. Chem. Soc.* 72, 909 (1950).

2. R. Livingston, L. Thompson and M.V. Ramarao, *ibid.* 74, 1073 (1952).

3. R. Livingston, *Quart. Rev. (London)* 14, 174 (1960).

4. D. Mauzerall, *J. Amer. Chem. Soc.* 82, 1832 (1960).

5. D. Mauzerall, *ibid.* 84, 2437 (1962).

6. G.R. Seely and K. Talmadge, *Photochem. Photobiol.* 3, 195 (1964).

7. A.N. Sidorov *in* "Elementary Photoprocesses in Molecules (B.S. Neoporent, ed.), p. 201. Consultants Bureau, (Plenum Press), New York, 1968.

8. D.G. Whitten, J.C. Yau and F.A. Carroll, *J. Amer. Chem. Soc.* 93, 2291 (1971).

9. D.G. Whitten, I.G. Lopp and P.D. Wildes, *ibid.* 90, 7196 (1968).

10. I.G. Lopp, R.W. Hendren, P.D. Wildes and D.G. Whitten, *ibid.* 92, 6440 (1970).

11. J.K. Roy and D.G. Whitten, *ibid.* 93, 7093 (1971).

12. J.K. Roy and D.G. Whitten, *ibid.* 94, 7162 (1972).

13. J.K. Roy, F.A. Carroll and D.G. Whitten, submitted for publication.

14. D.G. Whitten, P.D. Wildes and C.A. DeRosier, *J. Amer. Chem. Soc.* 94, 7811 (1972).

15. Chloro compounds have been found previously to quench excited triplets of naphthalene and other aromatics (16).

16. S. Ander, H. Blume, G. Heinrich and D. Schulte-Frohlinde, *Chem. Commun.* 745 (1968).

17. J.-H. Fuhrhop and D. Mauzerall, *J. Amer. Chem. Soc.* *91*, 4174 (1969).

18. J.H. Fuhrhop, *Structure and Bonding 18*, 1 (1974).

19. R.A. White and G. Tollin, *J. Amer. Chem. Soc. 89*, 1253 (1967).

20. B.J. Hales and J.R. Bolton, *ibid. 94*, 3314 (1972).

21. J.N. Demas and A.W. Adamson, *ibid. 95*, 5159 (1973).

22. C.R. Bock, T.J. Meyer and D.G. Whitten, submitted for publication.

23. A. Weller, *Pure Appl. Chem. 16*, 115 (1968) and references therein.

24. H. Knibbe, D. Rehm and A. Weller, *Ber. Bunsenges. Phys. Chem. 72*, 257 (1968).

25. D.A. Labianca, G.N. Taylor and G.S. Hammond, *J. Amer. Chem. Soc. 94*, 3679 (1972) and references cited therein.

26. N. Mataga and O. Tanimoto, *Theoret. Chim. Acta, (Berl.) 15*, 111 (1969).

27. H. Knibbe, D. Rehm and A. Weller, *Ber. Bunsenges. Phys. Chem. 73*, 839 (1969).

28. D. Rehm and A. Weller, *Israel J. Chem. 8*, 259 (1970).

29. F.A. Carroll, M.T. McCall and G.S. Hammond, *J. Amer. Chem. Soc. 95*, 315 (1973).

30. K.A. Zachariasse, private communication.

31. This equation ignores the "U" term which cannot easily be evaluated and thus probably represents lower limits for the charge-transfer state energy. The value for hexane solution (30) differs only slightly, as

$$E_{(S^+, Q^-, \text{ hexane})} = E_{1/2(S/S^+)} - E_{1/2(Q^-/Q)} + 0.13$$

$$\pm \ 0.10 \ eV.$$

32. N.J. Turro, "Molecular Photochemistry." p. 86. W.A. Benjamin, Inc., New York, 1967.

33. The ΔS* values measured here are similar to those observed in prophyrin dimerization (34,35) and suggest that in these complexes, as in the dimers, there is some regiospecificity.

34. J.H. Fuhrhop, P. Wasser, D. Riesner and D. Mauzerall, *J. Amer. Chem. Soc.* <u>94</u>, 7996 (1972).

35. K.A. Zachariasse and D.G. Whitten, *Chem. Phys. Letters* <u>22</u>, 527 (1973).

36. On theoretical grounds, as well as from experimental evidence, the contribution of excitation resonance to binding energies for triplet exciplexes is indicated to be very small (37).

37. W. Klöpffer *in* "Organic Molecular Photophysics." (J.B. Birks, ed.), Vol. 1, p. 397. J. Wiley and Sons, Ltd., London, 1973.

38. Since CT interactions are certainly important, it is perhaps surprising that no "CT" bands are detectable in the transient spectra. Possible alternative explanation to those advanced below could be that the spectra "accidentally" resemble those of the uncomplexed donor ( or that the CT bands are relatively weak or hidden in an inaccessible region of the spectra. In this regard it i interesting to note that ground-state porphyrin-*sym*-trinitrobenzene complexes have spectra not dramatically altered from those of uncomplexed porphyrins (40).

39. J.N. Murrell, "The Theory of the Electronic Spectra of Organic Molecules." p. 275. Methuen & Co., London, 1963 and references cited therein.

40. M. Gouterman and P.E. Stevenson, *J. Chem. Phys.* <u>37</u>, 2266 (1962).

41. C.A. Thayer and J.T. Yardley, *ibid.* <u>57</u>, 3992 (1972).

# EXCIPLEXES IN CHEMILUMINESCENT RADICAL ION RECOMBINATION

## KLAAS ZACHARIASSE

*Max-Planck-Institut fur biophysikalische Chemie,*

*Abteilung Spektroskopie, Göttingen, Germany*

## INTRODUCTION

Normally exciplexes, or hetero-excimers, $^1(A^-D^+)$ are formed by intermolecular interaction of an electronically excited species, *e.g.* an electron acceptor $^1A^*$, with a non-excited electron donor molecule D (or vice versa),

$$^1A^* + D \longrightarrow {}^1(A^-D^+) \tag{1}$$

However, chemiluminescent radical ion recombination (1) offers an alternative route for the formation and study of exciplexes. In this method the radical ions $A^-$ and $D^+$ are prepared chemically (2), and not electrochemically as in electrogenerated chemiluminescence,(3), although no fundamental difference between the two methods exists. As solvents, ethers such as tetrahydrofuran (THF) and 1,2-dimethoxyethane (DME) are used.

Two mechanisms for the formation of exciplexes by way of radical ion recombination have been described (1b, 1g, 4), *i.e.* direct formation from the radical ions (reaction [2]) and mixed triplet-triplet annihilation (reaction [3])

$$^2A^- + {}^2D^+ \longrightarrow ({}^2A^- \ldots {}^2D^+) \longrightarrow {}^1(A^-D^+) \qquad [2]$$

$$^2A^- + {}^2D^+ \longrightarrow ({}^2A^- \ldots {}^2D^+) \begin{array}{l} \nearrow {}^3A^* \ldots D \\ \searrow A \ldots {}^3D^* \end{array} \Bigg\} \; {}^3A^* + {}^3D^* \longrightarrow {}^1(A^-D^+)$$

$$[3]$$

The ion pair $({}^2A^- \ldots {}^2D^+)$ can best be formulated in a negative way: it is not the pair of ions $^2A^- + {}^2D^+$ far apart without any interaction, and also not the definite configuration of $A^-$ relative to $D^+$ found in the exciplex. A system $A^-/D^+$ where both direct formation and mixed triplet-triplet annihilation operate is bitolyl$^-$/TMPD$^+$ (see figure 1 and table 1) where TMPD = N,N,N'N'-tetramethyl-$p$-phenylenediamine.

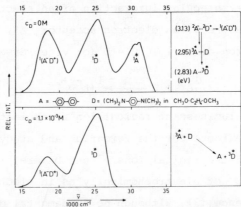

Fig. 1 *Chemiluminescence spectra (uncorrected, recorded with a RCA 1P28 photomultiplier) at room temperature with bitolyl radical anions (ca. 2×10$^{-4}$M) and Wurster's Blue perchlorate (TMPD$^+$ClO$_4^-$) in dimethoxyethane. Upper spectrum: no TMPD added. Lower spectrum: 1.1×10$^{-3}$M TMPD added to the radical anion solution.*

# TABLE 1

Data for bitolyl⁻/TMPD⁺ and 9-methylanthracene⁻/TPTA⁺ in tetrahydrofuran (THF)

| A | D | $\Delta G(A^- \ldots D^+)$ [a] | $\Delta E(^3A^*)$ | $\Delta E(^3D^*)$ | Chemiluminescence emitter | $h\nu_c^{max}$ (Solvent) [b] |
|---|---|---|---|---|---|---|
| Bitolyl | TMPD [c] | 3.13 | 2.95 | 2.83 | $^1(A^-D^+), ^1A^*, ^1D^*$ | 2.28 (THF) |
| 9-Methylanthracene | TPTA [d] | 2.92 | 1.81 | 2.96 | $^1(A^-D^+), ^1A^*$ | 2.43 (THF) |

TMPD: $E_{\frac{1}{2}}(D/D^+) = +0.16\ V$ vs S.C.E.; TPTA: $E_{\frac{1}{2}}(D/D^+) = +0.74\ V$ vs S.C.E.

All energy data are in eV.

[a] $\Delta G(A^- \ldots D^+)_{THF} = E_{\frac{1}{2}}(D/D^+) - E_{\frac{1}{2}}(A^-/A) + 0.20\ eV$ (1,4)

[b] Energy of the maximum of the exciplex emission, not corrected for the nonlinear sensitivity of the photomultiplier (RCA 1P28).

[c] TMPD = N,N,N',N'-tetramethyl-p-phenylenediamine

[d] TPTA = tri-p-tolylamine

Here the simultaneous operation of both mechanisms could be demonstrated *via* triplet-triplet energy transfer (1g, 4). Although complete quenching of the higher triplet (bitolyl) by the molecule having the lower triplet (TMPD) could be achieved, the exciplex emission was not completely quenched.

Of course, mechanism [3], mixed triplet-triplet annihilation, can only operate when both triplets $^3A*$ and $^3D*$ are energetically accessible, and therefore produced (1,4), from the ion pair ($^2A^-$ ... $^2D^+$). For systems where only one or no triplet state can be produced, clearly mechanism [2], direct formation from the ions, is the only mechanism available. Examples of systems where locally excited states of neither A nor D are accessible from ($^2A^-$ ... $^2D^+$), but where exciplex are nevertheless formed, are given in table 2 and figure 2.

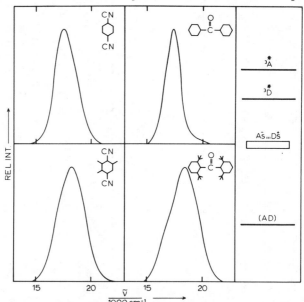

Fig. 2 *Chemiluminescence spectra of systems* $A^-$/TPTA$^+$, *TPTA=tri-p-tolylamine, where the free energy of the ion pair,* $\Delta G(A^-$ ... $D^+)$, *lies below the energies of the trip states of A and D. The structural formulae of A are giv in the spectra. (See table 2).*

TABLE 2

Data for radical ion recombination reactions for systems $A^-/D^+$ with $D$ = tri-p-tolylamine (TPTA) in methyltetrahydrofuran (MTHF), tetrahydrofuran (THF) and 1,2-dimethoxyethane (DME)

| A | $\Delta G(A^- \ldots D^+)^a$ | $\Delta E(^3A^*)$ | $\Delta E(^1A^*)$ | $h\nu_c^{max\,b}$ | Approximate quantum yield (Einstein/mole anion) |
|---|---|---|---|---|---|
| 2,6,2',6'-Tetraisopropyl-benzophenone | >2.65 | >3.01 | | 2.16 (MTHF) | $5 \times 10^{-5}$ |
| Benzophenone | 2.65 | 3.01 | 3.22 | 2.03 (THF) | $1 \times 10^{-4}$ |
| 2,3,6-Trimethyl-1,4-dicyano-benzene | 2.64 | 3.00 | >4.28 | 2.17 (DME) | $1 \times 10^{-4}$ |
| 1,4-Dicyanobenzene | 2.49 | 3.16 | 4.28 | 2.05 (THF) | $7 \times 10^{-4}$ |

$E_{1/2}(D/D^+) = +0.74$ V vs S.C.E.; $\Delta E(^3D^*) = 2.96$ eV; $\Delta E(^1D^*) = 3.51$ eV. All energy data are in eV.

a  $\Delta G(A^- \ldots D^+) = E_{1/2}(D/D^+) - E_{1/2}(A^-/A) + 0.20$ eV(1, 4)

b  $h\nu_c^{max}$: the energy of the maximum of the exciplex emission, not corrected for the nonlinear response of the photomultiplier (RCA 1P28).

Exciplex formation by way of radical ion recombination can be a very efficient process. Quantum yields of exciplex emission of $5 \times 10^{-2}$ photons/anion $A^-$ have been measured (1c) e.g. for anthracene$^-$/tri-$p$-tolylamine$^+$ (TPTA$^+$) in THF at room temperature. This means that, because of the large value of Avogadro's number, $3 \times 10^{17}$ photons of exciplex emission can be produced during the course of a chemiluminescence experiment, where 100 ml of a $10^{-4}$ $M$ anion solution react with the radical cation in approximately 100 seconds.

Differentiating between the two different mechanisms in exciplex formation can be accomplished with an external magnetic field (4). The presence of an external magnetic field is known to have an influence on the intensity of fluorescence brought about by triplet-triplet annihilation in molecular crystals (5) and in solution (6). The first magnetic field effect on electrochemiluminescent systems was reported by Faulkner and Bard (7).

Application of an external magnetic field of 1400 $\pm$ 100 gauss to bitolyl$^-$/TMPD$^+$ in THF, where the simultaneous operation of mechanisms [2] and [3] has been demonstrated, causes a reproducible change in the intensity ratio of the three emission bands in the chemiluminescence spectrum, which are ascribed to $^1(A^-D^+)$, $^1D^*$ and $^1A^*$, respectively; see figure 3 upper part. These excited states are produced, at least in part in the case of $^1(A^-D^+)$, by mixed or normal triplet-triplet annihilation. The intensity ratio $I_{1(A^-D^+)}/I_{(^1D^*)}$ increases by 13% on application of the field. Only relative changes in the chemiluminescence intensities can be studied because of the flow method (4) used in the radical ion recombination.

On the contrary, no such change in the chemiluminescence spectrum with magnetic field could be observed for a system

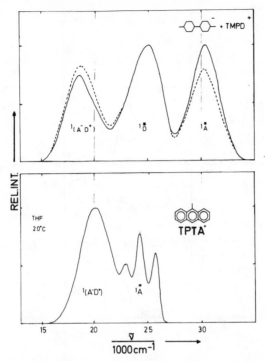

Fig. 3  Chemiluminescence spectra of bitolyl$^-$/TMPD$^+$
in 1,2-dimethoxyethane (upper half) and 9-methyl-
anthracene$^-$/TPTA$^+$ in tetrahydrofuran (lower half).
The dashed line in the upper figure is the spectrum
with an external magnetic field of approximately
1400 gauss.  The two spectra have been arbitrarily
adjusted so that the height of the $^1$D* peak is the
same in both cases.  For 9-methylanthracene$^-$/TPTA$^+$,
an influence of the magnetic field could not be
observed.  (See table 1).

such as 9-methylanthracene$^-$/TPTA$^+$ in THF (figure 2, lower
part), where exciplex formation does not involve triplet states
as precursors (mechanism [2], (1d).  (See table 1).  These
results corroborate and illustrate the operation of the two
different mechanisms of exciplex formation described above.

EXCIPLEXES IN RADICAL ION RECOMBINATION

Exciplex emission can be observed in many A$^-$/D$^+$ systems

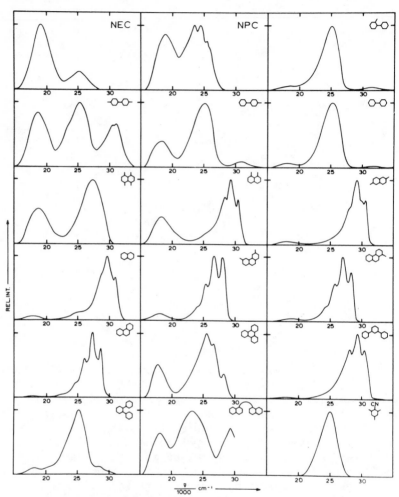

Fig. 4 Chemiluminescence spectra in tetrahydrofuran
or 1,2-dimethoxyethane resulting from recombination
reactions between Wurster's Blue perchlorate and
radical anions $A^-$. The valence bond dashes in the
structural formulae of A stand for methyl substituents
NEC = N-ethylcarbazole, NPC = N-phenylcarbazole. The
hetero-excimer emission is the structureless band
below 20,000 $cm^{-1}$. The spectrum in the middle of the
lowest row is for 1,3-di(1-naphthyl)propane. With
2,4,6-trimethylbenzonitrile only a very weak exciplex
emission could be observed, which is not visible in the
diagram. The spectra have not been corrected for the
nonlinear sensitivity of the photomultiplier (RCA 1P28

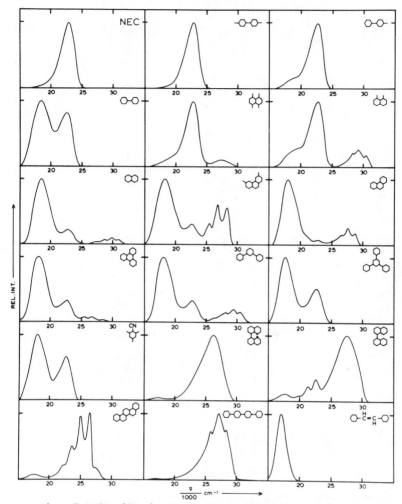

*Fig. 5  Chemiluminescence spectra at room temperature for systems $A^-/D^+$ where D is tris(p-dimethylamino-phenyl)amine (TPDA). The solvents used are 2-methyl-tetrahydrofuran, tetrahydrofuran or 1,2-dimethoxy-ethane. The valence bond dashes in the structural formulae of A stand for methyl substituents. For N-ethylcarbazole (NEC) an exciplex emission only is observed by lowering the temperature. With 1,1'-binaphthyl, but not with 8,8'-dimethyl-1,1'-binaphthyl, a fluorescence from perylene is found, apart from the emissions from $^1(A^-D^+)$ and $^1A^*$. The spectra have not been corrected for the nonlinear sensitivity of the photomultiplier (RCA 1P28).*

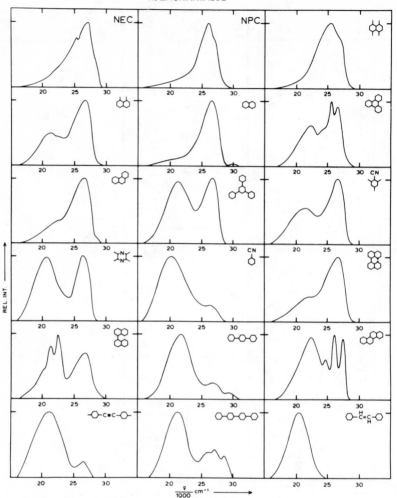

*Fig. 6  Chemiluminescence spectra at room temperature
of systems $A^-/D^+$ where $D$ = tri-p-tolylamine (TPTA), in
MTHF, THF or DME.  The valence bond dashes in the
structural formulae of A stand for methyl substituents
NEC = N-ethylcarbazole, NPC = N-phenylcarbazole.  The
structureless emissions around 21,000 $cm^{-1}$ originate
from $^1(A^-D^+)$.  The $^1D*$ emission has a maximum at
27,600 $cm^{-1}$;  other emissions originate from $^1A*$.  With
1,1'-binaphthyl, but not with 8,8'-dimethyl-1,1'-
binaphthyl, a fluorescence from perylene is observed
apart from the emissions from $^1(A^-D^+)$, $^1D*$ and $^1A*$.
The spectra have not been corrected for the nonlinear
sensitivity of the photomultiplier (RCA 1P28).*

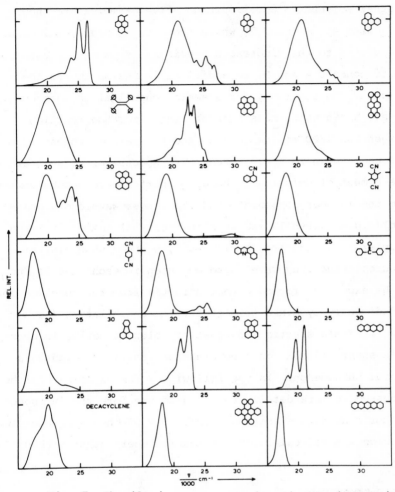

Fig. 7  Chemiluminescence spectra at room temperature
of systems $A^-/D^+$, where $D$ = TPTA, in MTHF, THF or
DME.  The valence bond dashes in the structural
formulae of A stand for methyl substituents.  The
structureless emissions around 20,000 cm$^{-1}$ originate
from $^1(A^-D^+)$;  in many cases an emission from $^1A*$ (but
in no case from $^1D*$) is observed.  The spectra have
not been corrected for the nonlinear sensitivity of
the photomultiplier (RCA 1P28).

in ether solvents. Some representative examples are shown in figures 4, 5, 6 and 7, where there is in nearly all cases, in addition to the fluorescence emission from one or both of the components A and D, an additional emission due to the exciplex. In figure 4, for example, chemiluminescence spectra are shown which have been observed from the reaction between the TMPD radical cation and the radical anions of the compounds shown in the diagram. The spectra are characterize by a number of emissions, those to shorter wavelengths coming from the monomer components and the longer wavelength emission (e.g. between 18 and 20kK in figure 4) coming from the exciplex. The spectra in figures 5, 6 and 7 show similar features; the long wavelength emission in each case is due to the exciplex, formed either directly from the reacting ion or indirectly by mixed triplet-triplet annihilation (see above). These spectra show that exciplex formation in chemiluminescent radical ion recombination reaction is indeed a general phenomenon. In the following a few examples will be discussed in more detail, to bring to the fore the influence of steric hindrance and the possibility of observing exciplex emission via radical ion recombination where other methods fail.

STERIC HINDRANCE

Excimer formation can be totally precluded by the presence of bulky substituents in a molecule. Thus, concentration quenching and excimer formation is observed for 9,10-dimethylanthracene, but not for 9,10-diphenylanthracene (8), where the phenyl groups form an angle of over $60^\circ$ with the anthracene plane (9,13).

Another case where the influence of steric hindrance is apparent has been observed in the dimer formation of zinc porphyrins (10). An influence of steric hindrance on fluorescence quenching reactions of a series of substituted anthracenes with N,N-diethylaniline in nonpolar solvents, as represented by reaction [1], could not be unambiguously observed, however (11).

The failure to detect exciplex emission for the nonplanar (helical) molecule 3,4-benzophenanthrene with N,N-dimethyl-aniline in cyclohexane has been explained by way of steric hindrance (12). Nevertheless, exciplex formation can in fact be readily observed for this system (4). In figure 8, the chemiluminescence spectrum of the system 3,4-benzophenanthrene$^-$/tris(p-dimethylaminophenyl)amine$^+$ in THF at room temperature

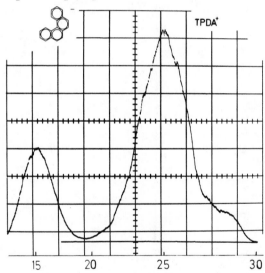

*Fig. 8 Chemiluminescence spectrum of the system 3,4-benzophenanthrene$^-$/TPDA$^+$ in THF at room temperature. The spectrum is a photograph taken directly from the oscillograph screen of the rapid scanning spectrofluorimeter, with an EMI 9658R photomultiplier. The horizontal axis gives the wavenumbers in 1000 cm$^{-1}$.*

is shown. An emission from the exciplex $^1(A^-D^+)$ is clearly present, $h\nu_c^{max}$ = 14,850 cm$^{-1}$, (see table 3). The emission with the maximum at 25,100 cm$^{-1}$ is the fluorescence from

TABLE 3

*Data for 3,4-benzophenanthrene⁻/TPDA⁺ in tetrahydrofuran*

| A | D | $\Delta G(A^- \ldots D^+)$ [a] | $\Delta E(^3A^*)$ | $\Delta E(^3D^*)$ | $\Delta E(^1A^*)$ | $\Delta E(^1D^*)$ | $h\nu_c^{max}$ [b] | Chemiluminescence emitter |
|---|---|---|---|---|---|---|---|---|
| 3,4-Benzophenanthrene | TPDA | 2.55 | 2.47 | 2.63 | 3.35 | 2.99 | 1.84 | $^1(A^- D^+)$, $^1A^*$ |

All energy data are in eV.

[a] $\Delta G(A^- \ldots D^+)_{THF} = E_{\frac{1}{2}}(D/D^+) - E_{\frac{1}{2}}(A^-/A)+0.20$ eV (1, 4); $-E_{\frac{1}{2}}(A^-/A) = 2.27$ V vs S.C.E.; $E_{\frac{1}{2}}(D/D^+) = +0.08$ V vs S.C.E.

[b] Energy of the maximum of the exciplex emission, not corrected for the nonlinear sensivity of the photomultiplier EMI 9658R.

3,4-benzophenanthrene, produced by triplet-triplet annihilation. The chemiluminescence spectrum in figure 8 is a photograph taken directly from the oscillograph screen of the rapid scanning spectrofluorimeter (14).

In this discussion on exciplexes in radical ion recombination, a few systems will be discussed where the radical anions are derived from molecules having bulky substituents: 9,10-diphenylanthracene, 9,10-di(1-naphthyl)anthracene and 9,9'-bianthryl. These molecules have a nonplanar structure, even as radical anions (13). Tri-$p$-tolylammonium perchlorate (TPTA$^+$ ClO$_4^-$), which itself also has a nonplanar structure, was used as the radical cation. A propellor-like configuration of TPTA$^+$ has been established, with the phenyl rings twisted out of the plane containing the central nitrogen atom and the methyl substituents (15). (Of course, radical ion recombination is not the only route to study these sterically hindered systems, if low solubility and/or short lifetime (see below) do not hinder the observation of exciplex formation by way of photoexcitation, reaction [1]).

As can be seen from figure 9, exciplex emission can easily be observed for these systems. The chemiluminescence spectra consist of emissions from $^1$(A$^-$D$^+$) and $^1$A*. An emission from $^1$D* does not occur, as is expected, because of the energy deficiency $\Delta G$(A$^-$ ... D$^+$) < $\Delta E$($^3$D*), $\Delta E$($^1$D*), (see table 4). The exciplex is formed directly from the ions (mechanism [2]), as mixed triplet-triplet annihilation cannot occur, since only one triplet state, $^3$A*, is generated from (A$^-$ ... D$^+$).

It can be concluded from the above results that, in contrast to the case of excimers, exciplex emission is not strongly disturbed by steric hindrance exerted by bulky substituents in the interacting molecules.

TABLE 4

*Systems $A^-/D^+$ in tetrahydrofuran ($\varepsilon=7.4$); $D$ = tri-p-tolylamine, $\Delta E(^3D*) = 2.96$ eV;*

*$\Delta E(^1D*) = 3.51$ eV; $E_{\frac{1}{2}}(D/D^+) = 0.74$ V vs S.C.E. $A$ is a sterically hindered molecule.*

| A | $-E_{\frac{1}{2}}(A^-/A)$ (V vs S.C.E.) | $\Delta G(A^- \dots D^+)$ [a] | $\Delta E(^3A*)$ | $\Delta E(^1A*)$ | Chemiluminescence emitter | $h\nu_c^{max}$ [b] |
|---|---|---|---|---|---|---|
| 9,10-Diphenyl-anthracene | 1.92 | 2.86 | 1.77 | 3.16 | $^1(A^-D^+)$, $^1A*$ | 2.37 |
| 9,10-Di(1-naphthyl)-anthracene | 1.90 | 2.84 | 1.77 | 3.16 | $^1(A^-D^+)$, $^1A*$ | 2.34 |
| 9,9-Bianthryl | 1.97 | 2.91 | 1.81 | 3.19 | $^1(A^-D^+)$, $^1A*$ | 2.34 |

All energy data are in eV.

[a] calculated according to the empirical (1,2,4) formula

$\Delta G(A^- \dots D^+)_{THF} = E_{\frac{1}{2}}(D/D^+) - E_{\frac{1}{2}}(A^-/A) + 0.20$ eV.

[b] energy of the maximum of the exciplex emission, not corrected for the nonlinear sensitivity of the photomultiplier (RCA 1P28).

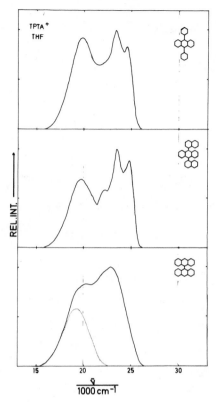

*Fig. 9   Chemiluminescence spectra of systems $A^-/TPTA^+$
in tetrahydrofuran at room temperature.   A is, from
top to bottom, 9,10-diphenylanthracene, 9,10-di(1-
naphthyl)anthracene and 9,9'-bianthryl.   The dashed
curve in the lowest spectrum is the exciplex emission.
(See table 4).   The spectra have not been corrected
for the nonlinear sensitivity of the photomultipler
(EMI 9658R).*

The factors determining the excimer stability are thought
to be governed by the intermolecular overlap between the
π-electron wavefunctions of the aromatic hydrocarbons forming
the excimer.   In exciplexes, by far the largest part of the
stabilization energy is contributed by the coulombic energy
of the interaction between the negatively charged $A^-$ and the
positively charged $D^+$.   This coulombic energy can easily be
determined to be of the order of 3 eV, in a point-charge model

calculated *via* Hückel orbitals, $A^-$ and $D^+$ lying in parallel planes 3.0 Å apart (29, 4).

The experimental results presented above, *i.e.* the very large sensitivity towards steric hindrance for excimers on the one hand, and the much smaller sensitivity for exciplexes on the other, could lead to the conclusion that the coulombic energy term $C(A^-D^+)$ is much less sensitive towards steric hindrance than is the intermolecular overlap.

CONDITIONS FOR OBSERVING EXCIPLEX EMISSION

What are, in general, the optimal conditions for observing a exciplex $^1(A^-D^+)$ *via* the (usual) reaction [1]? First, the primarily excited species, *e.g.* $^1A^*$, should have a long fluorescence lifetime and a large extinction coefficient as the excitation wavelength, enabling the production of an as large as possible stationary concentration of the primarily excited species, without the necessity of having to go to large concentrations ($>10^{-3}M$) for A. The use of such relatively large concentrations for the primarily excited species can in some cases lead to complications in the observation of exciplex emission, because of photophysical (*e.g.* 9,10-dimethylanthracene, triphenylamines) or photochemical reactions (anthracene). Second, the primarily non-excited species should be present in sufficiently large concentration to enable diffusional interception of the excited species during its lifetime.

All the factors outlined above can be deduced from an equation, valid for nonpolar solvents, based on the well-known Stern-Volmer relation (16, 17).

$$\frac{I'}{I} = \frac{\phi'}{\phi} \cdot k_q \tau_o \, [M] \qquad [4]$$

Here I' and I are the intensities and $\phi$ and $\phi'$ the quantum yields of the exciplex and monomer fluorescence, respectively. $k_q$ is the reaction rate constant of exciplex formation (reaction [1]), $\tau_o$ is the lifetime of the monomer fluorescence when [M], the concentration of the nonexcited species, is zero. Equation [4] applies to cases where there is no back reaction from the exciplex to , for example, $^1A* + D$.

For the reliable observation of exciplexes, the inequality I'/I > 0.05 should apply. This means that for a primarily excited molecule $^1A*$ with a lifetime of 2 nsec, a concentration of D larger than $2.5 \times 10^{-3}M$ is needed, with the assumption $\phi' = \phi$ and with $k_q = 1.0 \times 10^{10}$ 1 mole$^{-1}$ sec$^{-1}$ (16, 18). If $\tau_o$ is shorter, then the concentration of D must be higher. Conversely, if the solubility of D does not allow such a high concentration, $\tau_o$ must be larger. If these conditions cannot be satisfied, there will be little chance of observing exciplex formation.

However, the difficulties outlined above, caused by a short lifetime of $^1A*$ and a low solubility of D, are of no importance in the study of exciplex formation *via* radical ion recombination. Here, the solubility of the radical ions to be studied is the determining factor. In ether solvents such as THF, the radical anion concentration easily reaches the $10^{-5}M$ value where chemiluminescence can be readily observed.

An example of a system A/D which is difficult to study by way of photoexcitation because of the short lifetime of $^1A*$ is benzophenone/TPTA. As can be seen from figure 2, exciplex formation can easily be observed in this case *via* radical ion recombination.

POLYPHENYLS

As an example of a series of systems which cannot easily be studied by electron donor-acceptor interaction in the excited state (reaction [1]) because of low solubility, a series of $p$-polyphenyls has been selected.

The solubility of the higher polyphenyls is very slight, even in toluene (19) (see table 5). The solubility in other solvents such as the ethers used here, is even considerably smaller (20).

Exciplex formation $via$ chemiluminescent radical ion recombination has been studied for systems consisting of the radical cation of tris($p$-dimethylaminophenyl)amine (TPDA) and radical anions derived from the polyphenyls, biphenyl to $p$-sexiphenyl, in THF. The chemiluminescence spectra of the polyphenyl series are depicted in figure 10, $cf.$ table 5. They consist in all cases of a fluorescence from $^1A*$ (the polyphenyl) and from the exciplex $^1(A^-D^+)$. On the basis of the above results and backed by the large number of $A^-/D^+$ systems studied (4), deductions can be made as to properties of the polyphenyls, which are not easily accessible by other means.

First, consider the reduction potentials of $p$-quinquiphenyl and $p$-sexiphenyl. As is well known (21), a correlation exists between the reduction potential $E_{\frac{1}{2}}(A^-/A)$ and the energy of the lowest unoccupied Hückel-MO. On this basis, the reduction potentials of $p$-quinquiphenyl and $p$-sexiphenyl can be obtained by graphic extrapolation, using the measured values of $E_{\frac{1}{2}}(A^-A)$ for biphenyl, $p$-terphenyl and $p$-quaterphenyl (22); the data are given in table 4. This theoretical prediction can now be checked using the well-established

TABLE 5

Data for chemiluminescent systems in tetrahydrofuran with

$D = tris(p\text{-}dimethylaminophenyl)amine$

| A | $-E_{1/2}(A^-/A)$ (V vs S.C.E.) | $\Delta G(A^- \cdots D^+)$[b] | $\Delta E(^3A^*)$ | $\Delta E(^1A^*)$ | Chemiluminescence emitter | $h\nu_c^{max}$ (THF)[c] | Solubility of A in Toluene (15) | Approximate total quantum yield [cf.(1c)] |
|---|---|---|---|---|---|---|---|---|
| Biphenyl | 2.65 | 2.93 | 3.01 | 4.13 | $^1(A^-D^+), ^1D^*$ | 2.19[d] | | $3 \times 10^{-3}$ |
| p-Terphenyl | 2.31 | 2.59 | 2.55 | 3.99 | $^1(A^-D^+), ^1A^*$ | 1.96[e] | $3.2 \times 10^{-2}$ M | $1 \times 10^{-3}$ |
| p-Quaterphenyl | 2.17 | 2.45 | <2.45 | 3.72 | $^1(A^-D^+), ^1A^*$ | 1.88[e] | $3.9 \times 10^{-4}$ M | $2 \times 10^{-3}$ |
| p-Quinquiphenyl | (2.11)[a] | 2.39 | <2.39 | 3.59 | $^1(A^-D^+), ^1A^*$ | 1.84[e] | $7.9 \times 10^{-6}$ M | $4 \times 10^{-4}$ |
| p-Sexiphenyl | (2.05)[a] | 2.33 | <2.33 | 3.46 | $^1(A^-D^+), ^1A^*$ | 1.76[e] | | $1 \times 10^{-4}$ |

$E_{1/2}(D/D^+) = +0.08$ V vs S.C.E.; $\Delta E(^3D^*) = 2.63$ eV; $\Delta E(^1D^*) = 2.99$ eV. All energy data are in eV.

a   extrapolation based on the relation between $E_{1/2}(A^-/A)$ and the energy of the lowest vacant Huckel MO (17)

b   $\Delta G(A^- \cdots D^+)_{THF} = E_{1/2}(D/D^+) - E_{1/2}(A^-/A) + 0.20$ eV(1, 4)

c   energy of the maximum of the exciplex emission, not corrected for the nonlinear sensitivity of the photomultiplier, RCA 1P28 (d) and EMI 9658R (e).

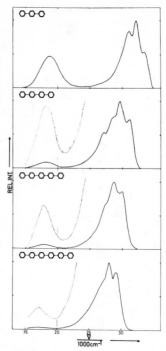

Fig. 10  Chemiluminescence spectra of systems $A^-/TPDA^+$
in tetrahydrofuran, where A is, from top to bottom,
p-terphenyl, p-quaterphenyl p-quinquiphenyl and
p-sexiphenyl. (See table 5.)  The spectra have not
been corrected for the nonlinear sensitivity of the
photomultiplier (EMI 9658R).

relation between the maximum of the exciplex emission ($h\nu_c^{max}$)
and the reduction potential of a series of electron acceptors
with one electron donor (16, 23).  A good agreement between
the value of $h\nu_c^{max}$ and the reduction potentials is found,
table 5.  Second, the appearance of  the strong fluorescence
of the polyphenyls in these energy deficient [i.e. $\Delta E(^1A^*) >$
$\Delta G(A^- \ldots D^+)$] systems, can only be brought about via triplet-
triplet annihilation.  This means that $\Delta G(A^- \ldots D^+) >$
$\Delta E(^3A^*)$, enabling one to give an upper limit to the energy
of the triplet states of p-quaterphenyl,  p-quinquiphenyl and

TABLE 6

*Data for chemiluminescent systems $A^-/D^+$ in tetrahydrofuran, with $A$ = chrysene*

| D | $E_{1/2}(D/D^+)$ (V vs S.C.E.) | $\Delta G(A^- \ldots D^+)$[a] | $\Delta E(^3D^*)$ | $\Delta E^1(D^*)$ | Chemilumi-nescence emitter | $h\nu_c^{max}$[b] |
|---|---|---|---|---|---|---|
| TPDA | + 0.08 | 2.58 | 2.63 | 2.99 | $^1(A^-D^+)$, $^1A^*$ | 1.98 |
| TMPD | + 0.16 | 2.66 | 2.83 | 3.52 | $^1(A^-D^+)$, $^1A^*$ | 1.93 |

$E_{1/2}(A^-/A) = -2.30\ V$ vs S.C.E.; $\Delta E(^3A^*) = 2.49\ eV$; $\Delta E(^1A^*) = 3.43\ eV$. Highly purified chrysene (du Pont) was used (0.0020% 2,3-benzocarbazole, 0.035% sulphur).

[a] $\Delta G(A^- \ldots D^+)_{THF} = E_{1/2}(D/D^+) - E_{1/2}(A^-/A) + 0.20\ eV$ (1, 4).

[b] energy of the maximum of the exciplex emission, not corrected for the nonlinear sensitivity of the photomultiplier (EMI 9658R).

and $p$-sexiphenyl, which are not easily accessible because of low solubility and small intersystem crossing yield. As a lower limit to the triplet state energy, of course, one has the condition $\Delta E(^3A*) > \frac{1}{2} \Delta E(^1A*)$, because of the operation of triplet-triplet annihilation.

EXCIPLEX INTENSITY

From an inspection of the chemiluminescence spectra of the polyphenyls (figure 10) and the approximate chemilumi-nescence quantum yields (table 5), it is seen that the intensity of the exciplex emission diminishes considerably from $p$-terphenyl to $p$-sexiphenyl. At the same time the energy corresponding to the uncorrected maximum of the exci-plex emission decreases from 1.96 to 1.74 eV. This means that the energy difference between the exciplex state $^1(A^-D^+)$ and the destabilized Franck-Condon state $^1(AD)$, reached just after emission, becomes smaller with increasing number of phenyl groups in the polyphenyl series. That the intensity of exciplex emission sharply decreases as the energy differ-ence between $^1(A^-D^+)$ and $^1(AD)$ becomes smaller than 2.0 eV is a generally observed phenomenon. It seems to be caused by the enhancement of the radiationless processes starting from the exciplex (4). In an extension of the theory of radiationless processes occuring in molecules such as aromatic hydrocarbons, a situation has been discussed where a relatively large change in configuration occurs as a result of the processes involved in the radiationless transition (24, 26), leading to an intersection of the potential energy curves describing the initial and the final state. The term "strong coupling limit" was introduced for this case. The processes starting from the initially prepared exciplex $^1(A^-D^+)$ clearly involve a relatively large change in configuration. The strong coupling limit therefore should apply to exciplexes as well as to excimers and EDA complexes, as was recognized by Siebrand for excimers (27) and by Freed and Jortner for EDA complexes (25).

For the probability W of the nonradiative decay in the strong coupling limit, Levich (28) and Freed and Jortner (25) derived an expression of the form

$$W = B \exp (-E_A/C) \qquad [5]$$

The factor B contains (25) the matrix element of the interaction operator H' between the zeroth order states $^1(A^-D^+)$ and $^1(AD)$. The "activation energy" $E_A$ is directly related to the energy difference $\Delta E$ between $^1(A^-D^+)$ and $^1(AD)$ (24). This means that W is directly proportional to the magnitude of the matrix element $< {}^1(A^-D^+) \, |H'| \, {}^1(AD)>$ and inversely proportional to the energy difference $\Delta E$. Therefore

$$W \sim \frac{<{}^1(A^-D^+) \, |H'| \, {}^1(AD)>}{E^1(A^-D^+) - E^1(AD)} \sim \frac{F.S(a',d)}{E^1(A^-D^+) - E^1(AD)} \qquad [6]$$

where the matrix element of the interaction operator H' between the zeroth order state $^1(A^-D^+)$ and $^1(AD)$, can be described (29, 4) as a product of a factor F and an intermolecular overlap intergral S(a', d) between the lowest unoccupied molecular orbital of the acceptor a', and the highest occupied orbital of the donor d.

That not only the energy above the ground state but also the matrix element $<{}^1(A^-D^+) \, |H'| \, {}^1(AD) >$ is an important factor determining the intensity of the exciplex emission can be seen from a comparison of the chemiluminescence spectra of chrysene$^-$/TPDA$^+$ and chrysene$^-$/TMPD$^+$, figure 11, cf. table 6. Here the chrysene emission $^1A^*$ has approximately the same quantum yield for both systems. The energy difference

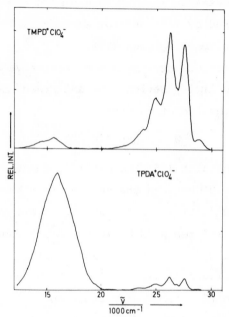

*Fig. 11  Chemiluminescence spectra of chrysene⁻/TMPD⁺ (above) and chrysene⁻/TPDA⁺ (below) in tetrahydrofuran. (See table 6.) The spectra have not been corrected for the nonlinear sensitivity of the photomultiplier (EMI 9658R).*

$E^1(A^-D^+) - E^1(AD)$ has approximately the same value for both systems, because of the similar values of the oxidation potentials $E_{\frac{1}{2}}(D/D^+)$, + 0.08 $V$ and + 0.16 $V$ vs S.C.E. respectively.

That the exciplex emission is much stronger for chrysene⁻/TPDA⁺ than for chrysene⁻/TMPD⁺, can be caused by the fact that the intermolecular overlap intergral $S(a',d)$, and therefore the matrix element $< {}^1(A^-D^+)|H'|{}^1(AD) >$ generally has a smaller value for the larger system A/TPDA than for A/TMPD (4).

One other factor influencing the intensity of exciplex

emission in chemiluminescent radical ion recombination is the yield of formation of the exciplex from the ion pair $(A^- \ldots D^+)$. This also depends on matrix elements such as $<^1(A^-D^+)|H'|^1(AD)>$, and has been discussed in previous papers (1f, 1g).

Measurements of lifetimes and quantum yields of exciplex emissions will be necessary in order to be able to describe accurately the phenomena presented above.

ACKNOWLEDGEMENT

Many thanks are due to Dr. G.A. Sloan, du Pont and Nemours, Wilmington, for making available highly purified chrysene. Fruitful discussions with Prof. Dr. A. Weller are gratefully acknowledged.

## REFERENCES

1. A. Weller and K. Zachariasse, a. *J. Chem. Phys.* _46_, 4984 (1967); b. *in* "Molecular Luminescence." (E.C. Lim, ed.), p. 895. Benjamin, New York, 1969. c. *Chem. Phys. Letters* _10_, 424 (1971); d. *ibid.* _10_, 590 (1971); e. *in* "Chemiluminescence and Bioluminescence." (M.J. Cormier, D.M. Hercules and J. Lee, eds.) p. 169. Plenum, New York, 1973; f. *ibid.* p. 181; g. *in* "Luminescence of Crystals, Molecules, and Solutions" (F. Williams, ed.) p. 219. Plenum, New York, 1973.

2. A. Weller and K. Zachariasse, *Chem. Phys. Letters* _10_, 197 (1971).

3. C.P. Keszthelyi and A.J. Bard, *Chem. Phys. Letters* _24_, 300 (1974) and earlier papers cited therein.

4. K. Zachariasse, Ph.D. Thesis, Free University, Amsterdam (1972).

5. R.E. Merrifield, *J. Chem. Phys.* _48_, 4318 (1968).

6. D. Wyrsch and H. Labhart, *Chem. Phys. Letters* _8_, 217 (1971).

7. L.R. Faulkner and A.J. Bard, *J. Amer. Chem. Soc.* _91_, 209 6495, 6497 (1969).

8. J.B. Birks , "Photophysics of Aromatic Molecules." Wiley-Interscience, New York, 1970.

9. B. Tinland, *Theor. Chim. Acta* _11_, 385 (1968).

10. K.A. Zachariasse and D.G. Whitten, *Chem. Phys. Letters* _22_, 527 (1973).

11. G. Rippen and K. Zachariasse, unpublished results.

12. M.A.F. Tavares, *Trans. Faraday Soc.* _66_, 2431 (1970).

13. L.S. Marcoux, A. Lomax and A.J. Bard, *J. Amer. Chem. Soc.* _92_, 243 (1970).

14. D. Rehm, Z. *Naturforsch.* *25A,* 1442 (1970).

15. D.W.A. Sharp, *J. Chem. Soc.* 4804 (1957).

16. H. Knibbe, Ph.D. Thesis, Free University, Amsterdam (1969).

17. H. Knibbe, D. Rehm and A. Weller, *Ber. Bunsenges.* *Phys. Chem.* *73,* 839 (1969).

18. H. Knibbe, D. Rehm and A. Weller, *ibid.* *72,* 257 (1968).

19. H.O. Wirth, F.U. Herrmann, G. Herrmann and W. Kern, *Mol. Cryst.* *4,* 321 (1968).

20. Handbook of Chemistry and Physics, (S.M. Elby, ed.), Chemical Rubber Publishing Company, Cleveland, Ohio.

21. A. Streitwieser, "Molecular Orbital Theory." Wiley, New York, 1961.

22. A.L. Allred and L.W. Bush, *J. Phys. Chem.* *72,* 2238 (1968).

23. H. Knibbe, D. Rehm and A. Weller, *Z. Phys. Chem. N.F.* *56,* 95 (1967).

24. R. Englman and J. Jortner, *Mol. Phys.* *18,* 145 (1970).

25. K.F. Freed and J. Jortner, *J. Chem. Phys.* *52,* 6272 (1970).

26. R. Englman and B. Barnett, *J. Luminescence 3,* 37 (1970).

27. W. Siebrand, *J. Chem. Phys.* *46,* 440 (1967).

28. V.G. Levich *in* "Advances in Electrochemistry and Electrochemical Engineering." (P. Delahay and C.W. Tobias, eds.), Vol. 4, p. 249. Interscience, New York, 1966.

29. H. Beens, Ph.D. Thesis, Free University, Amsterdam (1970).

# EXCIPLEXES IN ELECTROGENERATED CHEMILUMINESCENCE

ALLEN J. BARD AND SU MOON PARK

*Department of Chemistry*

*The University of Texas at Austin*

*Austin, Texas  78712*

## INTRODUCTION

Electrogenerated chemiluminescence (ECL) arises from electron transfer reactions of oxidized and reduced species generated electrochemically which produce excited states and ultimately light.  A typical reaction sequence involving an electron acceptor (A) and a donor (D) is

$A + e \rightarrow A^{\cdot -}$ (reduction at cathode or during cathodic

pulse) [1]

$D - e \rightarrow D^{\cdot +}$ (oxidation at anode or during anodic

pulse) [2]

$A^{-} + D^{+} \rightarrow A^{*} + D$ or $A + D^{*}$ (production of excited

states of A or D in the reaction layer).  [3]

These excited states can be singlet or triplet states, depending upon the energetics of the annihilation reaction and the rate of formation of the various states;  the radiative and nonradiative processes following their production (emission, quenching, triplet-triplet annihilation) are the same as those observed in photogenerated excited states. A number of reviews on ECL have appeared (1-6) or will appear shortly (7,8) so that only a brief outline of the basic concepts will be presented here and this work will mainly deal with the intermediacy of dimeric excited states (excimers or exciplexes) in ECL processes.

ECL is usually carried out in fairly conductive solutions prepared by adding a soluble, nonelectroactive salt (typically tetra-$n$-butylammonium perchlorate, TBAP) to an aprotic solvent with a wide available potential range and moderate (>7) dielectric constant [$e.g.$ acetonitrile; dimethylformamide (DMF); methylene chloride; tetrahydrofuran (THF)]. The acceptor and donor molecules comprise a variety of compounds, including aromatic hydrocarbons ($e.g.$ A = D = 9,10-diphenylanthracene), heterocyclic molecules ($e.g.$ A = D = tetraphenylporphin; A = 2,5-diphenyl-1,3,4-oxadiazole, D = thianthrene) or metal chelates [$e.g.$ A = D = tris(bipyridyl)-ruthenium(II)]. One should also mention that similar electron-transfer chemiluminescent (CL) reactions can be carried out by reacting ethereal solutions of radical anions (formed by alkali metal reductions of aromatic hydrocarbons) with dissolved species from solid radical-cation salts (typically perchlorate salts of N,N,N',N'-tetramethyl-$p$-phenylenediamine (TMPD) or tri-$p$-tolyamine (TPTA) radical cations)(9, 10). Thus CL studies can utilize solvents of lower dielectric constant ($e.g.$ dimethoxyethane) and involve much lower ionic strength solutions, but are limited to very stable reactant systems.

Several possible routes are found in the steps leading from the redox reaction to the production of light. The simplest and the one that shows the greatest efficiency (in terms of photon production per redox event) is the direct formation of an excited singlet state in the redox reaction (the S-route). This occurs for example with A = D = 9,10-diphenylanthracene (DPA), as

$$DPA^- + DPA^+ \rightarrow {}^1DPA^* + DPA \qquad [4]$$

and requires that the energy of the redox step be greater than that of the singlet excited state produced. Another possible path (the T-route) involves production of a triplet state in the redox reaction followed by triplet-triplet annihilation to produce the emitting state. This sequence occurs with A = DPA and D = TMPD, as

$$A^- + D^+ \rightarrow {}^3A + D \qquad [5]$$

$$^3A + {}^3A \rightarrow {}^1A^* + A \qquad [6]$$

For this case quenching of the triplet species by paramagnetic species (e.g. $A^-$, $D^+$, $O_2$) and other, nonparamagnetic, quenchers (Q) occurs:

$$^3A + D^+ \rightarrow A + D^+ \qquad [7]$$

$$^3A + Q \rightarrow A + Q \qquad [8]$$

The study of the effect of magnetic fields of up to 8 kG on the emission from ECL cells has been especially useful in uncovering the participation of triplet species in the light-producing process (11 - 14). Investigations of the ECL and delayed fluorescence of a number of systems have shown that a magnetic field will affect the rate of reactions involving triplet-triplet annihilation (reaction [6]) or quenching of triplets by paramagnetic species (e.g. reaction [7]), but apparently will not affect the electron transfer reaction [3]. Magnetic field effects have provided a useful probe of ECL reactions involving dimeric excited states, as will be demonstrated below.

ECL SYSTEMS INVOLVING EXCIMERS

Chandross, Longworth and Visco (15) were the first to suggest the direct formation of excimers in the radical-ion annihilation reaction

$$A^{-} + A^{+} \rightarrow {}^{1}A_{2}^{*} \qquad [9]$$

when emission at wavelengths longer than that for ${}^{1}A^{*}$ was observed for several aromatic hydrocarbons. These results were clouded somewhat, especially for A = anthracene, when it was discovered that all of the long wavelength emission could be accounted for by formation of anthranol (by reaction of anthracene radical cation with traces of water in the DMF solvent) which is excited by energy transfer from excited singlet anthracene (16, 17). This possibility of longer wavelength emission in an ECL system arising from a decomposition product is always present, especially where electrochemical studies demonstrate that either $A^{-}$ or $D^{+}$ is not stable and the solution exhibits new fluorescence bands after extensive ECL electrolysis. Parker and Short (18) utilized the kinetic analysis of relative excimer and monomer emission intensity, frequently used in spectroscopic studies (19), in a study of the ECL of 9,10-dimethylanthracene and explained the difference between the observed ECL behavior and that in delayed fluorescence as evidence for direct formation of excimer. These results have been disputed, however, and again side products or excimer formed by triplet-triplet annihilation have been suggested as the source of the long wavelength emission (17).

Maloy and Bard (20) studied the steady state ECL of the system, A = pyrene, D = TMPD, at the rotating ring-disk

lectrode; the energetics of the electron transfer reaction
n this case allow only formation of triplet pyrene (reaction
5]). The ECL spectrum clearly showed both a monomer com-
onent and an excimer band which is more intense than that
een in prompt fluorescence. An analysis of the relative
xcimer to monomer intensities with pyrene concentration was
onsistent with a mechanism in which triplet-triplet annihi-
ation leads to both $^1A^*$ and $^1A_2^*$ (reaction [10]). The effect
f a

$$^3A + \,^3A \rightleftharpoons (triplet\text{-}triplet) \left\langle \begin{array}{c} ^1A^* \\ \Updownarrow \\ ^1A_2^* \end{array} \right. \qquad [10]$$

lagnetic field on the intensities of both monomer and excimer
emission resulting only from triplet-triplet annihilation
vould be expected to be the same. However,Wyrsch and Labhart
(21) reported that the effect of a magnetic field on the
lelayed fluorescence of monomer (excited singlet) 1,2-benz-
anthracene (BA) at temperatures of -70 to -170°C was different from that of the excimer and concluded that a mechanism
involving a common triplet-triplet annihilation process could
not be the main route to $^1BA^*$ and $^1(BA)_2^*$ under these conditions. The magnetic field effect on the delayed fluorescence
of 1,2-benzanthracene and pyrene at room temperature was
recently investigated by Tachikawa and Bard (22), who reported
identical field effects for both monomer and excimer emission
in both cases, in agreement with a mechanism involving a
common intermediate produced on triplet-triplet annihilation.
Moreover the ECL of the pyrene$^-$/TMPD$^+$ system also showed
monomer and excimer emissions identically affected by a
magnetic field, as shown in figure 1. Excimers produced by

triplet-triplet annihilation were also found in the CL studie
of Weller and Zachariasse (9,10), for example in the reactior
of 9,10-dimethylanthracene radical anion with solid $TMPD^+$
$ClO_4^-$ in THF solutions.

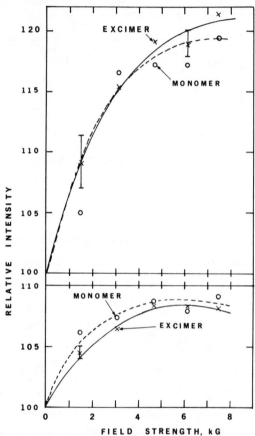

Fig. 1   Magnetic field effects on monomer and excimer
         ECL from
         (a) 1 mM pyrene, 1 mM TMPD and 0.1 M TBAP in DMF
         (b) 5 mM pyrene, 1 mM TMPD and 0.1 M TBAP in DMF
         (from ref. 22).

ECL SYSTEMS INVOLVING EXCIPLEXES

Exciplex emission can arise in ECL through three routes:

(i) the conventional singlet quenching route first
escribed by Weller and coworkers (23)

$$^1A^* + D \rightarrow {}^1(A^-D^+)^* \tag{11}$$

(ii) *via* triplet-triplet annihilation

$$^3A + {}^3D \rightarrow {}^1(A^-D^+)^* \tag{12}$$

(iii) on direct reaction of the radical ions

$$A^{\overline{\cdot}} + D^{\overset{+}{\cdot}} \rightarrow {}^1(A^-D^+)^* \tag{13}$$

ECL and CL studies are generally carried out at concen-
ration levels where exciplex production *via* reaction [11]
s not important (except where the radical annihilation
eaction is energetic enough to produce $^1A^*$ directly in the
icinity of D, so that [11] becomes an intermediate step in
13]). It is the direct formation path [13] of exciplex
ormation which is unique to ECL and CL processes. Weller
nd Zachariasse (9, 10) studied the reaction of a large
umber of radical anions with the perchlorate salts of the
adical cations of TMPD, TPTA, tri(*p*-dimethylaminophenyl)-
mine and tri-*p*-anisylamine in dimethoxyethane, THF and MTHF
olutions. In a number of cases, especially when $TPTA^+$ was
he oxidant, evidence for direct formation of exciplex *via*
13] was obtained. Since the lifetime of triplet TPTA is
ery short (*ca.* 50 nsec in ethanol at room temperature),
eaction [12] is a less likely source of exciplex in $TPTA^+/A^-$
eactions. Moreover exciplex emission was observed in cases
e.g. benzophenone$^-$/TPTA$^+$, see below) where neither triplet
tate was energetically accessible by the electron transfer
eaction. Several of these systems and other new ones have
ecently been investigated by ECL (24). Consider for example
he case where A = *trans*-stilbene ($E_{\frac{1}{2}}$(redn.) = -2.22 $V$ vs
g reference electrode, $E_S$ = 3.80 e$V$ and $E_T$ = 2.20 e$V$) and
= TPTA ($E_{\frac{1}{2}}$ (oxdn.) = +0.90 $V$ vs Ag reference electrode,

$E_S$=3.51 e$V$ and $E_T$=2.96 e$V$) in THF - 50 m$M$ TBAP solution ($E_S$ and $E_T$ are the energy levels of the first excited singlet and lowest triplet state respectively, and the half-wave potentials, $E_{\frac{1}{2}}$, were determined in the solvent vs a silver-wire reference electrode). The enthalpy of the redox reactio (*ca.* -3.02 e$V$) is insufficient to produce the singlet excited states. The triplet states of both species are very short-lived , so that the best explanation for the emission observed at 400 - 600 nm is direct exciplex formation in the redox step.

Another system reinvestigated by ECL was that of the 9-methylanthracene (MA)/TPTA system. Exciplex formation was observed in CL studies in THF solutions (9, 10) and the authors concluded, from a study of the relative exciplex to $^1$MA$^*$ emission as a function of solvent and temperature, that the exciplex is formed on direct reaction of MA$^-$ and TPTA$^+$ and that $^1$MA$^*$ is formed by thermal dissociation of the exciplex

$$^1(A^-D^+)^* \rightarrow {}^1A^* + D \qquad [14]$$

with triplet-triplet annihilation processes playing a negligible role at room temperature. In the ECL of a solution of 9-methylanthracene ($E_{\frac{1}{2}}$ (redn) = -1.94 $V$ vs S.C.E., $E_S$ = 3.20 e$V$, $E_T$ = 1.81 e$V$) and TPTA in THF solutions containing 0.2 $M$ TBAP, emission from excited singlet 9-methylanthracene and a longer wavelength emission at 520 nm are observed as shown in figure 2 (22). The effect of magnetic fields on both the 415 nm and 520 nm peaks is shown in figure 3. The intensities of both peaks are increased with increasing fields; however, the increase of the monomer (415 nm) peak is larger than that of the exciplex (520 nm). The large field effect on the 415 nm peak suggests appreciable

312

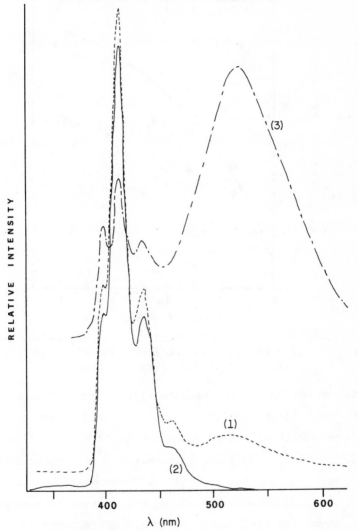

Fig. 2 (1) ECL spectrum of 1 mM MA, 1 mM TPTA and 0.1 M
TBAP in THF.
(2) Fluorescence spectrum of 1 mM MA, 1 mM TPTA
and 0.1 M TBAP in THF.
(3) Chemiluminescence spectrum from the reaction
of MA⁻ in THF with solid TPTA⁺Cl⁻ at 20°C
obtained by Weller and Zachariasse (9) (from
ref. 22).

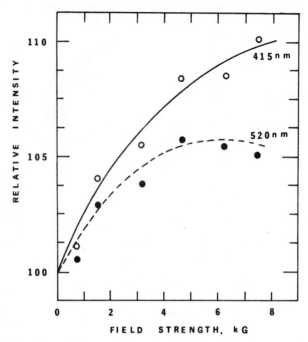

Fig. 3  Magnetic field effects on (O) 415 nm and (●)
520 nm ECL emission from 1 mM MA, 1 mM TPTA and
0.2 M TBAP in THF (from ref. 22)

formation of $^1MA^*$ by the triplet-triplet annihilation route
(reactions [5] and [6]),  The smaller field effect at 520 nm
could be attributed to emission by $^1MA_2^*$ formed by triplet-
triplet annihilation along with emission from directly-formed
exciplex.  One should note that differences in behavior are
often observed between the CL and ECL systems because of the
high supporting electrolyte (TBAP) concentrations usually
used in the latter studies (25).

Our most recent work (26) has been concerned with extending these studies on exciplex formation in ECL and especially
in elucidating the role of redox reaction energetics and
solvent in these reactions.  Only a brief summary of these
findings will be presented here;  a detailed report of this

work will be published elsewhere. The ECL resulting from
the reaction of TPTA$^+$ with the radical anions of a number
of acceptor molecules (benzophenone, acetophenone, dibenzoyl-
methane, naphthalene, 1,2-benzanthracene, 1,1,4,4-tetra-
phenyl-1,3-butadiene, 1,4-diphenyl-1,3-butadiene) in both
THF and acetonitrile show emissions at wavelengths longer
than those for $^1A^*$ and $^1D^*$; this emission can most reasonably
be attributed to exciplex formation. Similarly, the reaction
of naphthalene radical anion with radical cations of several
donor molecules (TPTA, triphenylamine, N,N'-diphenylbenzidine,
bis-1,8-(N,N-dimethylamino)naphthalene, tri-$p$-anisylamine) in
acetonitrile shows exciplex emission. Three typical cases
are discussed below.

## BENZOPHENONE$^-$/TPTA$^+$ (FIGURE 4)

For this system, the enthalpy for the redox reaction
(-2.59 eV in acetonitrile and -2.77 eV in THF) is smaller
than the triplet energies of both benzophenone ($E_T$ = 2.97 eV)
and TPTA ($E_T$ = 2.96 eV). The fairly bright, long wavelength
emission observed in both acetonitrile ($\lambda_{max}$ = 581 nm) and
THF ($\lambda_{max}$ = 594 nm) can only be attributed to exciplex formed
upon direct reaction of A$^-$ and D$^+$. Zachariasse (10) reached
a similar conclusion in a CL study with THF as a solvent.
While the exciplex emission intensity was lower in acetonitrile than in THF, the emission from an exciplex in this
polar solvent was clearly evident.

## NAPHTHALENE$^-$/TPTA$^+$ (FIGURE 5)

The naphthalene radical anion is fairly stable in
acetonitrile ($E_p$ (redn) = -2.66 V vs S.C.E.) and very stable
in THF ($E_p$ (redn) = -2.79 V vs S.C.E.) The reaction
enthalpy is -3.34 eV (acetonitrile) or -3.48 eV (THF),

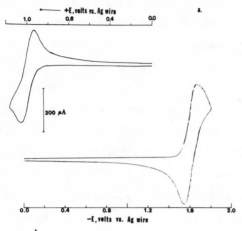

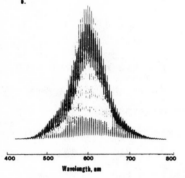

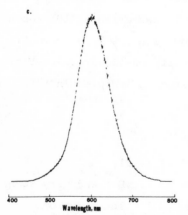

Fig. 4 _ECL of the benzo-
phenone$^-$/TPTA$^+$ system.
(a) Cyclic voltammogram of
2.2 mM TPTA, 2.5 mM benzophenone and 0.1 M TBAP in
THF at Pt electrode;  scan
rate was 200 mV/sec.
(b) ECL in above solution by
pulsing at single electrode
with pulse duration of 1 sec
(c) As in (b) with pulse
duration of 50 msec.

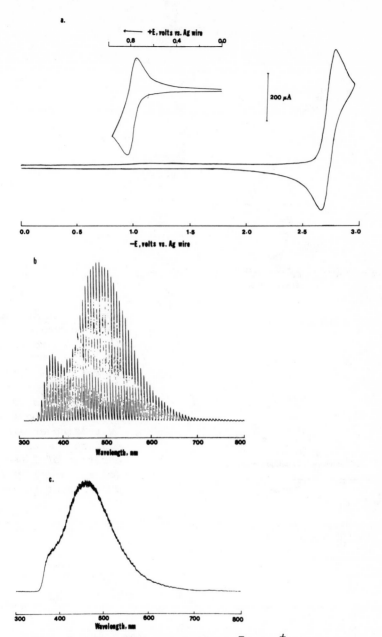

Fig. 5  ECL of the naphthalene⁻/TPTA⁺ system.
(a)  Cyclic voltammagram of 2.0 mM TPTA, 2.5 mM
naphthalene and 0.1 M TBAP in THF at Pt electrode;
scan rate was 200 mV/sec.
(b)  ECL in above solution by pulsing at single
electrode with pulse duration of 1 sec.
(c)  As in (b) with pulse duration of 50 msec.

317

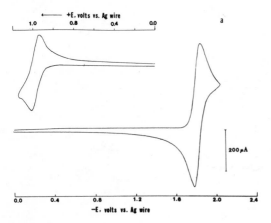

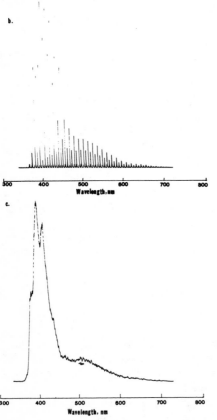

Fig. 6  ECL of the BA⁻ /
TPTA⁺ system.
(a) Cyclic voltammogram
of 1.3 mM TPTA, 2.3 mM
BA and 0.1 M TBAP in
THF at Pt electrode;
scan rate was 200 mV/sec
(b) ECL in above solution
by pulsing at single
electrode with pulse
duration of 1 sec.
(c) As in (b) with pulse
duration of 50 msec.

sufficient to produce the lowest triplet states of both naphthalene ($E_T$ = 2.63 e$V$) and TPTA. The observed ECL emission in both solvents consists of a band where TPTA fluoresces and a longer wavelength band ($\lambda_{max}$ (acetonitrile) = 476 nm; $\lambda_{max}$ (THF) = 467 nm) attributable to exciplex emission. Two paths are available to form the exciplex, mixed triplet-triplet annihilation [12] and direct formation [13]. The mixed triplet-triplet annihilation path is unlikely here however because of the previously mentioned short life-time of $^3$TPTA, as well as the absence of any naphthalene excited-singlet emission resulting from triplet-triplet anni-hilation of naphthalene triplets. The excited singlet of TPTA ($E_S$ = 3.51 e$v$) must then be produced by thermal activa-tion from the ion pair precursor state; a similar mechanism was invoked by Weller and Zachariasse (9, 10) for the CL in the 9-methylanthracene$^-$/TPTA$^+$ system. The singlet state of naphthalene ($E_S$ = 3.99 e$V$) clearly lies at too high an energy to be populated by this route. The ratio of exciplex to monomer emission was 1.96 in acetonitrile and 2.37 in THF.

## $,2$-BENZANTHRACENE$^-$/TPTA$^+$ (FIGURE 6)

The 1,2-benzanthracene (BA) radical anion is very stable in both acetonitrile ($E_p$ (redn) = -2.17 $V$ vs S.C.E.) and THF $E_p$(redn) = -2.24 $V$ vs S.C.E.). The ECL emission in acetoni-trile shows peaks at 391, 413 and 436 nm, corresponding to ,2-benzanthracene fluorescence, and a broad band at about 429 nm. Similar emission, with the long wavelength band at 411 nm, is observed in THF. The reaction enthalpy is 2.85 e$V$ in acetonitrile and 2.92 e$V$ in THF, sufficient to produce BA ($E_T$ = 2.08 e$V$) but not $^1$BA$^*$ ($E_S$ = 3.35 e$V$). Thus $^1$BA$^*$ is formed by triplet-triplet annihilation and some of the long wavelength emission might be attributed to 1,2-benzanthracene excimer (reaction [10]). However, the fact that this band shifts slightly when changing solvents and that the ratio of long wavelength to monomer emission is 0.071 in acetonitrile and 0.18 in THF lends support to at least some exciplex emission as well. The general behavior of the system appears to parallel that of the 9-methylanthracene$^-$/TPTA$^+$ system, in which magnetic field effect studies led to a similar mechanism (22).

## CRITERIA FOR ESTABLISHING EXCIPLEX EMISSION IN ECL

We might conclude by reviewing some of the criteria which allow long wavelength emission observed in ECL systems to be attributed to exciplex emission.

### ELECTROCHEMICAL STUDIES

The electrochemical (*e.g.* cyclic voltammetric) behavior allows one to establish the chemical stability of the reactant species (A$^-$ and D$^+$). Long wavelength emission observed in systems showing lack of chemical reversibility in either the reduction or oxidation peaks should be looked upon with suspicion, since reaction products may be responsible for it. The electrochemistry also allows the determination of peak potentials, and, from these, the reaction enthalpies.

### CORRELATIONS WITH ELECTRODE POTENTIALS

The exciplex energy is related to the ionization potentials of the donor molecules and the electron affinities of the acceptor molecules (27, 28). Since the ionization potential correlates with the electrode potential for oxidation of a molecule while the electron affinity correlates with the electrode potential for reduction (29), a linear correlation between E(A$^-$D$^+$), the exciplex emission maximum, and E$_{\frac{1}{2}}$ for reduction of A (for a given D with a series of acceptors) or between $E(A^- D^+)$ and $E_{\frac{1}{2}}$ for oxidation of D (for a given A with a series of donors) is expected (26). Such correlations have been observed for the exciplexes investigated here (figures 7 and 8).

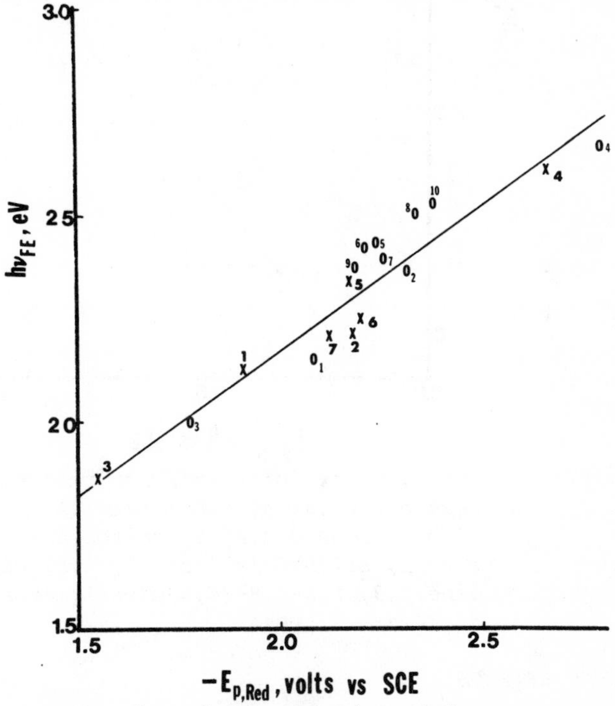

Fig. 7 *Energy of exciplex emission maximum vs re-*
*duction peak potentials of acceptor for ECL of*
*TPTA with various acceptors in (O) THF and (X)*
*acetonitrile solutions. Acceptors were 1,*
*benzophenone; 2, acetophenone; 3, dibenzoyl-*
*methane; 4, naphthalene; 5, 1,2-benzanthracene;*
*6, 1,1,4,4-tetraphenyl-1,3-butadiene; 7, 1,4-*
*diphenyl-1,3-butadiene; 8, trans-stilbene (24);*
*9, 9-methylanthracene (22); 10, 2,5-diphenyl-*
*1,3,4-oxadiazole (24).*

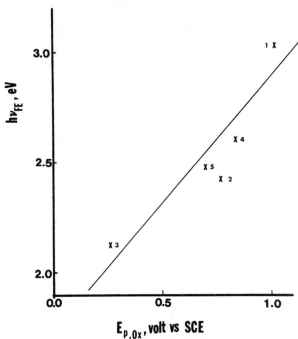

*Fig. 8   Energy of exciplex emission maximum vs oxidatic peak potential of donor for ECL of naphthalene with various donors in acetonitrile.   Donors were 1, triphenylamine;   2, N,N'-diphenylbenzidine;   3, bis-1,8-(N,N-dimethylamino)naphthaler. 4, TPTA;   5, TPAA.*

*SOLVENT EFFECTS*

Low dielectric constant solvents favor exciplex formatic so that the relative intensity of exciplex emission should decrease in proceeding from low to high dielectric constant solvents.   This effect was noted for the three systems described above.   Similarly the exciplex emission for the dibenzoylmethane$^-$/TPTA$^+$ system decreases in the order THF (7.6) > $CH_2Cl_2$ (9.08) > benzonitrile (25.5) > DMF (36.7) > acetonitrile (37.5);   no ECL emission is seen for this syste in dimethyl sulfoxide (46.6).   Moreover the exciplex emissic maximum should shift to shorter wavelengths in lower dielectric constant solvents (29);   the behavior of the benzophenone$^-$/TPTA$^+$ system is apparently anomalous in this respect. Note that exciplex emission is observed in acetonitrile in these studies, as opposed to failure to observe exciplexes in many spectroscopic studies in higher dielectric constant solvents.

## MAGNETIC FIELD EFFECTS

The relative magnetic field effect on monomer and longer wavelength emission is useful in uncovering the extent of participation of triplet-triplet annihilation and in distinguishing between excimers formed by triplet-triplet annihilation and directly-formed exciplexes.

## CONCLUSIONS

The formation of exciplexes in ECL processes affords an alternate route to these species. Indeed some exciplexes which are unavailable by photoexcitation paths (for example those involving singlet benzophenone, where photoproduced benzophenone singlets rapidly undergo intersystem crossing to the triplet state) can be produced by redox reactions. Since exciplexes are often invoked as intermediates in photochemical reactions, ECL offers the possibility of carrying out "photochemistry" without light (but rather by electrochemical excitation). Finally the demonstration of exciplexes as intermediates in the radical-ion annihilation process may have some bearing on the efficiency of photon emission in ECL processes. The radiationless deactivation of such intermediates may be the cause of low efficiencies in many ECL systems and perhaps is the reason why the ECL from rather sterically-hindered systems such as 9,10-diphenyl-anthracene, rubrene or tris(bipyridyl)ruthenium(II) is the most efficient (24).

ACKNOWLEDGMENT

Earlier contributions to the study of excimer and exciplex emission in ECL in this laboratory were made by C.P. Keszthelyi and H. Tachikawa. The support of this research by the U.S. Army Research Office - Durham is gratefully acknowledged.

REFERENCES

1.  T. Kuwana *in* "Electroanalytical Chemistry."
    (A.J. Bard, ed.), Vol. 1, Chap. 3. Marcel Dekker,
    New York, 1966.

2.  A.J. Bard, K.S.V. Santhanam, S.A. Cruser and
    L.R. Faulkner *in* "Fluorescence." (G.G. Guilbault, ed.),
    Chap. 14. Marcel Dekker, New York, 1967.

3.  A. Zweig, *Advances in Photochemistry* 6, 425 (1968).

4.  E.A. Chandross, *Trans. N.Y. Acad. Sci. Ser.* 2, 31,
    571 (1969).

5.  D.M. Hercules, *Accounts Chem. Res.* 2, 301 (1969).

6.  D.M. Hercules *in* "Physical Methods of Organic Chemistry."
    (A. Weissberger and B. Rossiter, eds.), 4th Ed., Part II.
    Academic Press, New York, 1971.

7.  A.J. Bard and L.R. Faulkner *in* "Creation and Detection
    of the Excited State" (W.R. Ware, ed.), Vol. 3. Marcel
    Dekker, New York

8.  T. Wilson and L.R. Faulkner *in* "Chemical Kinetics."
    (D.R. Herschbach, ed.), MTP Series in Physical Chemistry,
    (in preparation).

9.  A. Weller and K. Zachariasse, *J. Chem. Phys.* 46, 4984
    (1967); *Chem. Phys. Letters* 10, 197, 424, 590 (1971).

10. K.A. Zachariasse, Ph.D. Thesis, Vrije Universitet te
    Amsterdam (1972).

11. L.R. Faulkner and A.J. Bard, *J. Amer. Chem. Soc.* 91,
    209 (1969).

12. L.R. Faulkner, H. Tachikawa and A.J. Bard, *ibid.* 94,
    691 (1972).

13. H. Tachikawa and A.J. Bard, *Chem. Phys. Letters* 19,
    287 (1973).

14. N. Periasamy, S.J. Shah and K.S.V. Santhanam, *J. Chem. Phys.* *58*, 821 (1973).

15. E.A. Chandross, J.W. Longworth and R.E. Visco, *J. Amer. Chem. Soc.* *87*, 3259 (1965).

16. L.R. Faulkner and A.J. Bard, *ibid.* *90*, 6284 (1968).

17. T.C. Werner, J. Chang and D.M. Hercules, *ibid.* *92*, 763 (1970).

18. C.A. Parker and G.D. Short, *Trans. Faraday Soc.* *63*, 2618 (1967).

19. See *e.g.* C.A. Parker, *Spectrochim. Acta* *22*, 1677 (1960); J.B. Birks, B.N. Srinivasan and S.P. McGlynn, *J. Mol. Spectroscopy* *27*, 266 (1968); B. Stevens, *Chem. Phys. Letters* *3*, 233 (1969) and references cited therein.

20. J.T. Maloy and A.J. Bard, *J. Amer. Chem. Soc.* *93*, 5968 (1971). See also T. Kihara, M. Sukigara and K. Honda, *J. Electroanal. Chem.*, *47*, 161 (1973).

21. D. Wyrsch and H. Labhart, *Chem. Phys. Letters* *8*, 217 (1971)

22. H. Tachikawa and A.J. Bard, *ibid.* *26*, 568 (1974).

23. H. Leonhardt and A. Weller, *Ber. Bunsenges. Phys. Chem.* *67*, 791 (1963); A. Weller, *Pure Appl. Chem.* *16*, 115 (1968) and references cited therein.

24. C.P. Keszthelyi and A.J. Bard, *Chem. Phys. Letters* *24*, 300 (1974).

25. C.P. Keszthelyi, N.E. Tokel-Takvoryan, H. Tachikawa and A.J. Bard, *ibid.* *23*, 219 (1973).

26. S.M. Park and A.J. Bard, unpublished work

27. R.S. Mulliken and W.B. Person, "Molecular Complexes - A Lecture and Reprint Volume." Wiley-Interscience, New York, 1969 and references cited therein.

28. H. Knibbe, D. Rehm and A. Weller, *Z. Phys. Chem.*, *N.F.* *56*, 95 (1967).

# ELECTRON-TRANSFER REACTIONS IN MULTICOMPONENT SYSTEMS

S. FARID, S.E. HARTMAN AND T.R. EVANS

*Research Laboratories, Eastman Kodak Company*

*Rochester, New York   14650*

## INTRODUCTION

Exciplexes and excited charge-transfer complexes are known to dissociate in polar solvents into radical ion pairs (1). Few examples of dimerization of olefins *via* the radical cations so formed are known (2 - 5). For example, irradiation of an acetonitrile solution of 2,4,6-triphenyl-pyrylium fluoroborate (TPP) and 1,1-dimethylindene, A, led to cyclobutadimerization of the olefin (4).

The dimerization of phenyl vinyl ether, B, (6) was also found to proceed *via* a radical cation mechanism (5).

In an attempt to explore the scope and limitations of the reactions of such olefinic radical cations, we first investigated the possibility of their mixed additions.

Irradiation of an equimolar solution of A and B with TPP led to the formation of two isomers of a 1:1 adduct (AB) as the major product; small amounts of the dimers $A_2$ and $B_2$ were also formed.

The ratio of *endo:exo* adducts† from this reaction is 60:40. The formation of these adducts *via* a radical cation mechanism

---

† From the nmr spectra of the adducts not only were the structure and the stereochemistry established, but also a considerable distortion of the cyclobutane rings was evident.

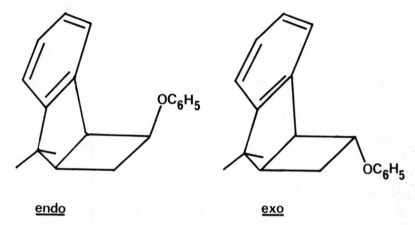

**endo**                                    **exo**

could also be achieved by excitation of ground-state charge-
transfer complexes of the reactants with different anhydrides.
For example, phthalic anhydride (PA) forms, with both A and
B, CT complexes that have weak end absorption at the 366 nm
Hg line. The optical density (OD) of an acetonitrile solution of 0.3 $M$ A and 0.1 $M$ PA at 366 nm through a 4 cm cell
is 0.26 which is 9.3 times that of the corresponding complex
with B. Excitation of an equimolar solution of both olefins
and PA at this wavelength leads, therefore, to the excited
complexes, $(PA...A)^*$ and $(PA...B)^*$, in the ratio 9.3:1. The
ratio of these and other complexes with different anhydrides
on excitation at 366 or 405 nm, together with the product
distribution of their reactions, is given in table 1.

It is interesting that a plot of the ratio of the adducts
AB to the dimer, $B_2$, vs the ratio of the excited complexes
gave a straight line. By extrapolation to $(X...A)^*/(X...B)^*$
= 0, a ratio of 12 for $AB/B_2$ was obtained. This corresponds
to the adduct:dimer ratio initiated by the formation of the
radical cation, $B^{+\cdot}$, in an equimolar solution of A and B. As
shown below, we reached the same conclusion from a concentra-
tion-dependency experiment.

The product distribution of the triplet-sensitized

$$\text{(PA)} \quad + \quad A \quad + \quad B \quad \underset{}{\overset{CH_3CN}{\rightleftharpoons}} \quad (PA\ldots A) + (PA\ldots B)$$

$$1 \; : \; 1$$

$$\downarrow h\nu$$

$$AB \quad + \quad B_2 \quad \longleftarrow \quad (PA\ldots A)^* + (PA\ldots B)^*$$

$$18 \; : \; 1 \qquad\qquad 9.3 \; : \; 1$$

TABLE 1

*Relative optical densities of CT complexes in equimolar solutions of A and B with different anhydrides (X) in $CH_3CN$ and the product distribution from the irradiated solutions*

| Anhydride (X) | $\lambda$ | $(OD)_{X\ldots A}/(OD)_{X\ldots B}$ | $AB/B_2$ |
|---|---|---|---|
| Tetrachlorophthalic anh. | 366 | 1.4 | 13.0 |
| Tetrachlorophthalic anh. | 405 | 5.4 | 15.3 |
| Phthalic anh. | 366 | 9.3 | 17.9 |
| Dimethylmaleic anh. | 366 | 10.0 | 18.8 |
| 4-Methylphthalic anh. | 366 | 11.9 | 20.6 |

reaction of A and B is distinctly different from that occurring *via* the radical cation. Almost equal amounts of the adducts AB and the dimer, $A_2$, are formed, and the ratio of *endo:exo* adducts is 45:55.

$$\text{A} + \text{B} \xrightarrow[\text{h}\nu,\ CH_3CN]{\substack{\text{Michler's ketone} \\ \text{or benzophenone}}} \text{AB} + \text{A}_2$$

$$1 \quad : \quad 1 \qquad\qquad\qquad\qquad\qquad 45 \quad : \quad 55$$

Whereas it is obvious that in the triplet-sensitized reaction only triplet-excited dimethylindene ($^3$A*) is formed, which reacts with A and B leading to dimer, $A_2$, and adduct formation, respectively, the mechanism of the photoinduced electron-transfer reaction is very complex.

## KINETICS AND MECHANISM

The following experiments were conducted to explore the interrelationship between electron-transfer and addition reactions of the different radical cations in this system.

### DIMERIZATION OF PHENYL VINYL ETHER

We first studied the kinetics of the photodimerization of B and determined several reaction constants involved in this process. When 9,10-dicyanoanthracene (DCA), was used as a sensitizer, the quantum yield of dimerization was determined as a function of the concentration of B. A plot of $\phi_{B_2}^{-1}$ vs $[B]^{-1}$ was linear, as shown in figure 1 with the slope = 0.081 and the intercept = 0.6. This intercept corresponds to a quantum yield of 1.7 at infinite concentration of B. Keeping the concentration of B constant at 0.42 $M$ and irradiating in the presence of increasing amounts of 1,5-dimethoxy-naphthalene (Q), which has a much lower oxidation potential than B, led to a change in $\phi_{B_2}^{-1}$ as a function of [Q] as shown in figure 2. This plot indicates the presence of two different intermediates involved in the dimerization reaction both of which can be intercepted by Q. These and other data from experiments to be described can be best rationalized in

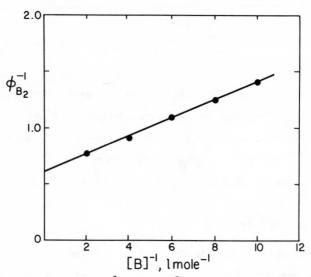

Fig. 1   Plot of $\phi_{B_2}^{-1}$ vs $[B]^{-1}$ from irradiations in $CH_3CN$
with DCA as a sensitizer at 405 nm.   Slope =
0.081 mole $l^{-1}$, intercept = 0.6.

terms of the following mechanism.

Scheme 1

$$^1S* + B \longrightarrow S\overset{-}{\cdot} + B\overset{+}{\cdot} \quad \text{(efficiency:f)}$$

$$B\overset{+}{\cdot} + X \xrightarrow{k_1} B + X\overset{+}{\cdot} \qquad [1]$$

$$B\overset{+}{\cdot} + Q \xrightarrow{k_2} B + Q\overset{+}{\cdot} \qquad [2]$$

$$B\overset{+}{\cdot} + B \xrightarrow{k_3} B_2\overset{+}{\cdot} \qquad [3]$$

$$B\overset{+}{\cdot} + Y \xrightarrow{k_4} B + Y\overset{+}{\cdot} \qquad [4]$$

$$B_2\overset{+}{\cdot} + X \xrightarrow{k_5} B_2 + X\overset{+}{\cdot} \qquad [5]$$

$$B_2\overset{+}{\cdot} + B \xrightarrow{k_6} B_2 + B\overset{+}{\cdot} \qquad [6]$$

$$B_2^{+\cdot} + Y \xrightarrow{\quad k_7 \quad} B_2 + Y^{+\cdot} \qquad [7]$$

$$B_2^{+\cdot} + Q \xrightarrow{\quad k_8 \quad} B_2 + Q^{+\cdot} \qquad [8]$$

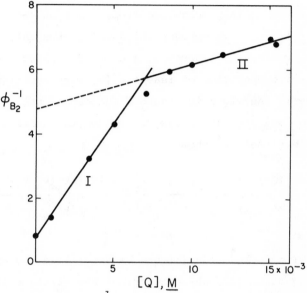

Fig. 2  Plot of $\phi_{B_2}^{-1}$ vs the concentration of 1,5-
dimethoxynaphthalene [Q] added to acetonitrile
solutions of 0.42 M phenyl vinyl ether, B, sensitized by DCA.  I; slope = 710 l mole$^{-1}$, intercept = 0.83:  II;  slope = 140 l mole$^{-1}$, intercept = 4.75.

In reactions [1] and [5], X can be $S^{-\cdot}$ or impurities in
the solution having a lower oxidation potential than both
B and $B_2$.  In reactions [4] and [7], Y represents impurities
in B having low oxidation potentials, i.e. [Y] = y [B], where
y is the molar fraction of the impurities in B.  Reaction [7]
is introduced to account for the fact that extrapolation to

infinite [B] in figure 1 does not lead to $\phi_{B_2} = \infty$.[†] Since
the oxidation potentials of B and $B_2$ are very similar (5), we
consider in the kinetic expressions that $k_1[X] = k_5[X] = k_x$,
$yk_4 = yk_7 = k_y$ and $k_2 = k_8 = k_q$.

The high efficiency of the quenching of the dimerization
at low [Q] (figure 2) is, according to scheme 1, due to interception of the chain-propagating step, reaction [6], by
reaction [8]. At high concentrations of Q, the chain propagation is practically suppressed and the quenching then
results from competition of reaction [2] with reaction [3].
The change in the slope in figure 2 indicates that $k_3 > k_6$.

The following equation was derived on the basis of the
reactions outlined in scheme 1.

$$\phi_{B_2}^{-1} = \frac{(k_x + k_q[Q])^2 + (k_x + k_q[Q])(k_3 + k_6 + 2k_y)[B] + k_y(k_3 + k_6 + k_y)[B]^2}{fk_3[B](k_x + k_q[Q] + (k_6 + k_y)[B])}$$

[1]

For [Q] = 0, equation [1] can be simplified and approximated to equation [2], which describes the straight line in
figure 1.

$$(\phi_{B_2}^{-1})_{[Q] = 0} \approx \frac{k_y(k_3 + k_6 + k_y)}{fk_3(k_6 + k_y)} + \frac{kx(k_3 + k_6 + 2k_y)}{fk_3(k_6 + k_y)}\frac{1}{[B]}$$

[2]

At constant [B], equation [1] can be written in the form of

[†] A reaction of $B_2^{+\cdot}$ with B leading to products (*e.g.* oligo-
mers) would be an alternative to reaction [7] to account for
$\phi_{B_2} \neq \infty$ at [B] $= \infty$.

equation [3], which at low [Q] is approximated to equation [4] and at high [Q] to equation [5].

$$\phi_{B_2}^{-1} = \frac{a + b[Q] + c[Q]^2}{d + e[Q]} \qquad [3]$$

$$\phi_{B_2}^{-1} \approx \frac{a}{d} + \frac{b}{d}[Q] \qquad [4]$$

$$\phi_{B_2}^{-1} \approx \frac{b}{e} + \frac{c}{e}[Q] \qquad [5]$$

$$b \approx k_q(k_3 + k_6 + 2k_y)[B]$$

$$c = k_q^2$$

$$d \approx fk_3(k_6 + k_y)[B]^2$$

$$e = fk_3 k_q [B]$$

Accordingly, equation [4] describes the initial line with higher slope in figure 2 and equation [5] that with lower slope.

To check the consistency of the proposed mechanism, we kept the concentration of Q constant at a high value $(1.5 \times 10^{-2} M)$ and varied the concentration of B. According to the data of figure 2, the chain propagation is negligible under these conditions. The change in $\phi_{B_2}$ will be described by equation [5], which can be written in form of equation [6].

$$(\phi_{B_2}^{-1})_{[Q]>10^{-2} \text{ const.}} \approx \frac{k_3 + k_6 + 2k_y}{fk_3} + \frac{k_q[Q]}{fk_3}\frac{1}{[B]} \qquad [6]$$

A plot of $\phi_{B_2}^{-1}$ vs $[B]^{-1}$ is shown in figure 3. The slope and intercept of this plot are interrelated with those in figure 2-II, as indicated from equations [5] and [6]. Within experimental error, these values were found to be in very good agreement.

335

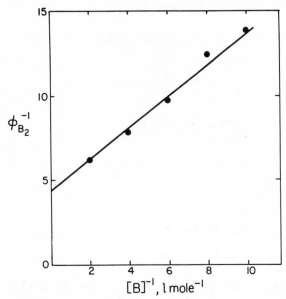

Fig. 3   Plot of $\phi_{B_2}^{-1}$ vs $[B]^{-1}$ from irradiations in presence of 1,5-dimethoxynaphthalene (1.5 × $10^{-2}$ M); slope = 0.92 mole $1^{-1}$, intercept = 4.4.

From the slopes and intercepts given in figures 1 - 3, the following reaction constants were determined assuming that $k_q = k_{diff} = 10^{10}$ 1 mole$^{-1}$ sec$^{-1}$, as justified on the basis that the oxidation potential of Q is much lower than that of B and $B_2$:

$k_3 = 6.3 \times 10^8$, $k_6 = 1.4 \times 10^8$, $k_y = 2 \times 10^7$ 1 mole$^{-1}$ sec$^{-1}$

$k_x = 2.7 \times 10^6$ sec$^{-1}$, and f = 0.27.

*REACTIONS OF PHENYL VINYL ETHER RADICAL CATION WITH*

*DIMETHYLINDENE*

The magnitude, direction and possible reversibility of electron transfer between the radical cations of A and B will

depend on the difference of their oxidation potentials; the
oxidation potentials (vs S.C.E. in $CH_3CN$) of A and B (5) were
found to be 1.68 and 1.75 eV. The oxidation of A is a
reversible process; that of B, however, is nonreversible.
Nevertheless, there are several other indications that A has
a lower oxidation potential than that of B: all the CT complexes of anhydrides with A absorb at longer wavelength than
those with B; the reaction constant for the fluorescence
quenching of DCA with A is 1.6 times that for B; and in the
mass spectrum of the adduct AB, the molecular ion cleaves
predominantly to the ion $A^{+\cdot} + B$ and not to $B^{+\cdot} + A$.

The reaction of A and B initiated by $B^{+\cdot}$ was studied by
irradiating solutions of high concentration of B and varying
low concentrations of A. Under these conditions, mainly the
radical cation $B^{+\cdot}$ is formed from the reaction with $^1S*$. A
plot of the ratio $AB/B_2$ vs $[A]/[B]$ gave a straight line
going through the origin (figure 4) with a slope of 12, which
corresponds to the adduct/dimer ratio produced in an equimolar
solution of A and B initiated by $B^{+\cdot}$. We reached the same
conclusion from the experiments with anhydrides mentioned
above.

To differentiate between the two following mechanisms,
reaction [9] or reactions [10] - [12], we carried out the
reactions in the presence of an electron donor functioning
as a quencher Q.

$$B^{+\cdot} + A \xrightarrow{\ k_9\ } AB^{+\cdot} \xrightarrow{\ +e^-\ } AB \qquad [9]$$

$$B^{+\cdot} + A \xrightarrow{\ k_{10}\ } B + A^{+\cdot} \qquad [10]$$

$$B_2^{+\cdot} + A \xrightarrow{\ k_{11}\ } B_2 + A^{+\cdot} \qquad [11]$$

$$A^{+\cdot} + B \xrightarrow{\ k_{12}\ } AB^{+\cdot} \xrightarrow{\ +e^-\ } AB \qquad [12]$$

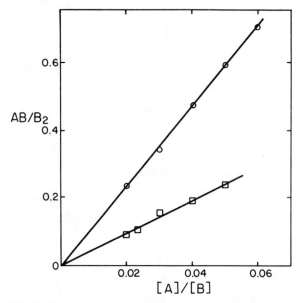

Fig. 4 *Plot of adduct:dimer ratio $AB/B_2$ vs [A]/[B] from irradiations of 0.3 M solutions of B at different concentrations of A. ○———○ ; in absence of Q, slope = 12: □———□; in presence of $10^{-3}$ M 1,5-dimethoxynaphthalene (Q), slope = 4.8.*

If AB is formed according to equation [9], a plot of $AB/B_2$ vs [A]/[B] in the presence of Q should have the same slope as that in absence of Q, but if AB is formed *via* equations [10] - [12] then a lower slope owing to interception of A $\overset{+}{\cdot}$ by Q should be obtained.

$$A \overset{+}{\cdot} + Q \longrightarrow A + Q \overset{+}{\cdot} \qquad [13]$$

This was found to be the case; with [Q] = $10^{-3}$ M the slope changed from 12 to 4.8 (figure 4).

Reactions [10] and [11] are analogous to reactions [2] and [8] of scheme 1, where Q is replaced by A. From a plot of $\phi_{B_2}^{-1}$ vs [A] at [B] = 0.3 M, we obtained a straight line with a slope of 100. When this value is compared with the slope of ~ 1000 obtained from the corresponding reaction in which A is replaced by 1,5-dimethoxynaphthalene (Q), a value of ~ $10^9$ l mole$^{-1}$ sec$^{-1}$ for $k_{11}$ and a similar value for $k_{10}$ are deduced. The fact that this value of $k_{11}$ is higher than that of $k_6$ for the corresponding reaction with B is consistent with this electron-transfer process since the oxidation potential of A is lower than that of B.

*REACTIONS OF THE RADICAL CATION OF DIMETHYLINDENE (A)*

The dimethylindene dimer, $A_2$, is a poorer quencher for the fluorescence of DCA than is A. The Stern-Volmer slopes for these quenchers are 71 and 178 l mole$^{-1}$, respectively. This indicates that $A_2$ has a higher oxidation potential than A and so would favor a chain mechanism. From a plot of $\phi_{A_2}^{-1}$ vs [A]$^{-1}$, an intercept very close to zero was obtained, which can be regarded as support for the chain mechanism (figure 5).

The slope of this plot is 2.3, compared with 0.08 for the corresponding reaction with B. This seems to be due to a small value for the reaction constant $k_{14}$. This was evident from the very large drop in $\phi_{A_2}$ in the presence of 1,5-dimethoxynaphthalene (Q), on the basis of which a value of ~ $10^6$ l mole$^{-1}$ sec$^{-1}$ is estimated for the bimolecular reaction constant $k_{14}$.

$$A^+ + A \xrightarrow{k_{14}} A_2^+ \qquad [14]$$

The reaction of A$^+$ with B was studied by keeping the concentration of A at a high constant value and varying B at low concentrations. The ratio of AB/$A_2$ was plotted vs [B]/[A]; a slope of 160 was obtained. Under the assumption that the reverse process in reactions [14] and [15] is of a minor magnitude, the reaction constant $k_{15}$ will be approximately two orders of magnitude higher than $k_{14}$, *i.e.*, $k_{15}$ is in the range of $10^8$ 1 mole$^{-1}$ sec$^{-1}$.

$$A \overset{+}{\cdot} + B \xrightarrow{\phantom{xx} k_{15} \phantom{xx}} AB \overset{+}{\cdot} \qquad [15]$$

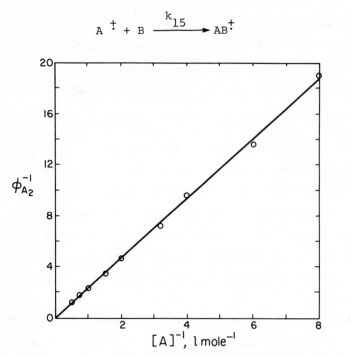

Fig. 5  *Plot of $\phi_{A_2}^{-1}$ vs $[A]^{-1}$ from irradiations in $CH_3CN$ with DCA as a sensitizer.  Slope = 2.3 mole $l^{-1}$.*

## COSENSITIZERS IN RADICAL CATION REACTIONS

We also obtained evidence that the endothermic reverse electron transfer step in equation [10] takes place.  On using 9-cyanoanthracene (CA) as a sensitizer for the dimerization of B, a very low quantum yield ($\phi_{B_2} = 3.3 \times 10^{-4}$ at $[B] = 0.3$ *M*) was measured.  The practically unmeasurable quenching of the fluorescence of CA by this olefin is in accordance with the low quantum yield.  The dimerization of A was also ineffective with this sensitizer ($\phi_{A_2} = 7.3 \times 10^{-3}$

at [A] = 0.3 $M$). In this case the fluorescence quenching of
the sensitizer was also inefficient but seemed to be higher
than that with B. An interesting observation was that a
0.3 $M$ solution in both A and B sensitized by CA led to the
adduct AB ($\phi_{AB}$ = 1.2 × 10$^{-2}$) and to the dimer B$_2$ at an
*increased* quantum yield ($\phi_{B_2}$ = 6.5 × 10$^{-4}$). This finding
can be rationalized in terms of scheme 2, in which the radical cation B$^{+\cdot}$ is formed in an electron-transfer reaction
from A$^{+\cdot}$. The formation of the latter species from the
reaction of A with the excited sensitizer is less endothermic than that leading to B$^{+\cdot}$ directly.

Scheme 2

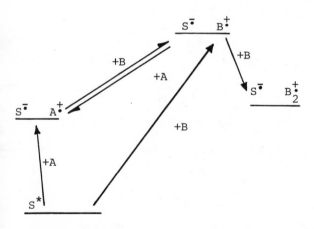

Strong support for this hypothesis was obtained by
replacing A in this reaction by other compounds of similar
oxidation potentials. The quantum yield, $\phi_{B_2}$, was found to
increase by a factor of up to 200 depending on the oxidation
potential and the concentration of the added compound
functioning as a cosensitizer. Phenanthrene, 1- and

2-methylnaphthalene, 2,3- and 2,6-dimethylnaphthalene and durene in concentrations up to 0.2 *M* were used as cosensitizers in these experiments. The same effect was observed in reactions sensitized by 9-fluorenone, which in the absence of a cosensitizer is an even poorer sensitizer than CA. The dimerization of A and the mixed addition of A to B, sensitized by CA or fluorenone, is also enhanced in the presence of such cosensitizers.

## REFERENCES

1.  K.H. Grellmann, A.R. Watkins and A. Weller, *J. Phys. Chem.* 76, 469 (1972); Y. Taniguchi and N. Mataga, *Chem. Phys. Letters* 13, 596 (1972).

2.  A. Ledwith, *Accounts Chem. Res.* 5, 133 (1972).

3.  R.A. Neunteufel and D.R. Arnold, *J. Amer. Chem. Soc.* 95, 4080 (1973).

4.  S. Farid and S.E. Shealer, *Chem. Commun.* 677 (1973).

5.  T.R. Evans, R. Wake and O. Jaenicke *in* "The Exciplex" (M. Gordon, W.R. Ware, P. de Mayo and D.R. Arnold, eds.), Academic Press, New York, 1975.

6.  S. Kuwata, Y. Shigemitsu and Y. Odaira, *J. Org. Chem.* 38, 3803 (1973).

SINGLET QUENCHING MECHANISMS - SENSITIZED

DIMERIZATION OF PHENYL VINYL ETHER

T.R. EVANS, R.W. WAKE   AND O. JAENICKE

Research Laboratories, Eastman Kodak Company

Rochester, New York  14650

INTRODUCTION

Photosensitization can be used to produce a wide variety
of reactive species.  In the early years of modern photochemistry, studies of triplet-state properties and reactions
were predominant.  It has recently been recognized that other
reactive species can also be produced by photosensitization
and attention is rapidly beginning to focus on their properties and reactions.  The major interest has been in the chemistry proceeding from both exciplexes and radical ions.

Although radical ions can be produced thermally, photochemical production often has distinct advantages, especially
in the case of radical cations.  Thermal generation of radical cations usually requires rather harsh conditions, which
may lead to multielectron transfer products rather than the
primary, one-electron transfer products.  In contrast to the
often acidic conditions for thermal generation, photosensitization can produce radical cations in neutral, acidic, or

345

even basic media. Furthermore under photochemical conditions, the radical ions are produced in dilute solution so that ion-ion interactions, often leading to dications, are improbable.

Finally, of special interest to chemists is the fact that photoinduced electron-transfer reactions often lead to products in a chain reaction (1-3). Although photoiniated free-radical chain reactions are common, photoiniated ionic chain reactions offer new possibilities for efficient polymer and organic syntheses (1-5). The chemistry of the olefinic radical cations $\underline{1}$, produced by photosensitization, has been investigated in a few cases.

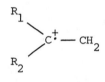

$$\underline{1}$$

When $\underline{1}$ is either a mono or disubstituted amino olefin, the radical cation undergoes dimerization to give cyclobutane products in a chain reaction (2). When $R_1 = R_2 = $ phenyl, dimerization is accompanied by rearrangement in a low quantum yield process (6). We were interested in the structure-reactivity relationships of substitued olefin-radical cations and particularly in what requirements are necessary to observe chain reactions. For our initial study we chose to examine the photosensitized reactions of phenyl vinyl ether $\underline{2}$. It has been reported (7) that $\underline{2}$ upon photochemical

$$C_6H_5OCH = CH_2$$

$$\underline{2}$$

sensitization with dimethyl terephthalate, methyl benzoate, or benzonitrile yields the dimers 3 and 4.  These authors stated

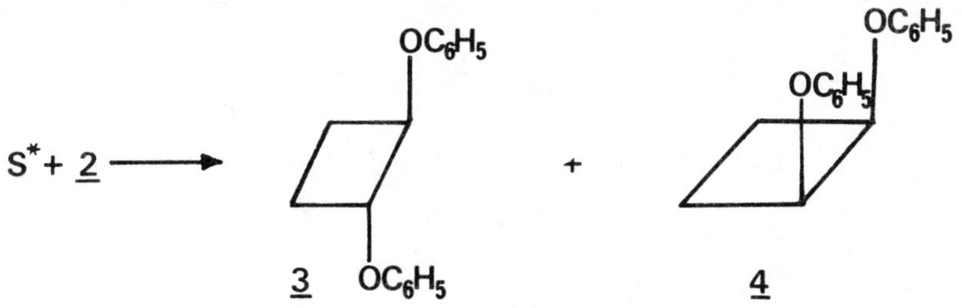

$$S^* + \underline{2} \longrightarrow \qquad \underline{3} \qquad + \qquad \underline{4}$$

that these dimers might be formed by some type of exciplex
mechanism, but their results were inconclusive.  Since no
quantum yields were reported, we could not tell whether this
reaction was a chain reaction or a low quantum yield process.
For these reasons we decided to reinvestigate the reaction.

RESULTS

    As reported, photosensitized reaction of 2 yields 3 and
4 as the major products when it is carried out in a polar
solvent.  Qualitative examinations indicated that 9,10-
dicyanoanthracene, 5, was better than any of the sensitizers
used by Kuwata et al.(7), and it was therefore used for all
of the studies reported in this paper.

    The fluorescence quenching efficiency of 5 by 2 and the
quantum yield for dimer formation are both extremely solvent
dependent (8).

    At room temperature there is a difference of nearly two
orders of magnitude between the fluorescence quenching
efficiency in polar vs  nonpolar solvent, as shown in table 1.
At room temperature, no exciplex is observed in any solvent
including methylcyclohexane;  in methylcyclohexane, at $-40^\circ$C,

347

the quenching efficiency increases and exciplex emission is observed (figure 1). Where $k_d$ is the diffusion-limited rate,

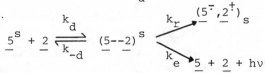

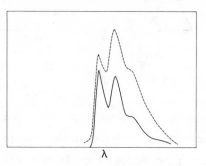

*Fig. 1  The emission from 9,10-dicyanoanthracene with (dotted line) and without (solid line) added phenyl vinyl ether. The fluorescence spectrum of the dicyanoanthracene has been normalized to indicate its fluorescence remaining after quenching with 0.35 M phenyl vinyl ether. In order to match the observed spectrum, a broad band with a maximum at approximately 470 nm must be added to the spectrum as shown. The maxima for the fluorescence of 9,10-dicyanoanthracene are at 424, 451 and 478 nm.*

$k_{-d}$ is the rate constant for dissociation of the exciplex, $k_r$ is the rate for ion-pair formation, and $k_e$ is the fluorescence rate constant for exciplex emission, we obtain the following expression for the observed rate of quenching

$$k_q = \frac{k_d (k_r + k_e)}{k_{-d} + k_r + k_e} \qquad [1]$$

The quenching constant $k_q$ is obtained from the usual Stern-Volmer equation $\phi_o/\phi = 1 + k_q T[Q]$. At room temperature $(k_{-d} + k_r) > k_e$ since no exciplex emission is observed, and we assume that $k_r > k_e$ in all solvents. With this assumption we can rearrange equation [1] to give

$$\frac{k_q}{k_d - k_q} = \frac{k_r}{k_{-d}} \qquad [2]$$

which will show linear behavior when $\ln k_q/k_d - k_q$ is plotted vs solvent polarity if the rate constant $k_r$ is polarity dependent and $k_{-d}$ is relatively independent of solvent. Such a plot is shown in figure 2, where the Dimroth polarity parameter is used (11); we feel it lends strong support to the idea that ionic species are the major products in the sensitized reaction of 2.

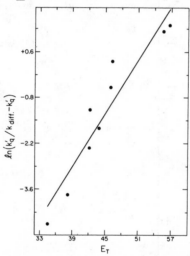

Fig. 2  The quenching efficiency of phenyl vinyl ether towards 9,10-dicyanoanthracene fluorescence as a function of solvent polarity.

These observations, the exciplex emission at $-40^{\circ}C$ in methylcyclohexane and the dependence of polarity on the fluorescence quenching at room temperature, are consistent with the formation of a weakly bound exciplex, which in non-polar solvents dissociates to $\underline{5}^S$ and $\underline{2}$ and in polar solvents forms the ion pair, figure 3.

In addition to the solvent effect, we also observe a salt

TABLE 1

Data for the fluorescence quenching of dicyanoanthracene by phenyl vinyl ether and dimer production from phenyl vinyl ether.

| Solvent | $E_T$ | Viscosity cp, $25°C$ | $K_{diff}$, $\times 10^{-10}$, $25°C$, l mole$^{-1}$ sec$^{-1}$ | $K_{sv}^e$ | $\phi([\underline{1}] = 0.2\,M)$ + dimer |
|---|---|---|---|---|---|
| Formamide | 56.6 | 3.3 | 0.2 | $23.5 \pm .5$ | |
| Methanol | 55.5 | 0.547 | 1.2 | $135 \pm 2$ | |
| Nitromethane | 46.3 | 0.620 | 1.07 | $90 \pm 2$ | 0.96 |
| Acetonitrile | 46.0 | 0.345 | 1.91 | $105 \pm 2$ | 0.73 |
| Dimethylformamide | 43.8 | 0.796 | 0.83 | $18.1 \pm 0.3$ | |
| Acetone | 42.2 | 0.316 | 2.09 | $73 \pm 3$ | |
| Benzonitrile | 42.0 | 1.24 | 0.53 | $6.9 \pm 0.3$ | |
| Ethyl acetate | 38.1 | 0.441 | 1.50 | $5.1 \pm 0.2$ | 0.08 |
| Benzene | 34.5 | 0.608 | 1.09 | $1.45 \pm 0.2$ | $<10^{-6}$ |

350

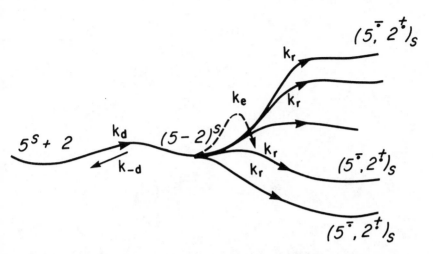

*Fig. 3 A potential energy diagram illustrating the dependence of fluorescence quenching ability upon the rate constant for electron transfer.*

effect on the quenching of 5 by 2 in a relatively nonpolar solvent. In dichloromethane, with 0.3 $M$ 2, 5 was quenched by the addition of tetrabutylammonium perchlorate. The Stern-Volmer plot was linear in tetrabutylammonium perchlorate concentrations with a quenching constant of 5.9. No quenching was observed in the absence of 2. Thus, in dichloromethane, the quenching efficiency of 2 doubles with the addition of 0.17 $M$ tetrabutylammonium perchlorate.

In acetonitrile, the dimers 3 and 4 are formed in a 60:40 ratio in a chain process with a limiting quantum yield of 1.5 (12). The dimers are not the only products formed in acetonitrile. In acetonitrile, and to a lesser extent in benzonitrile, the pyridine derivative 6 is formed along with phenol.

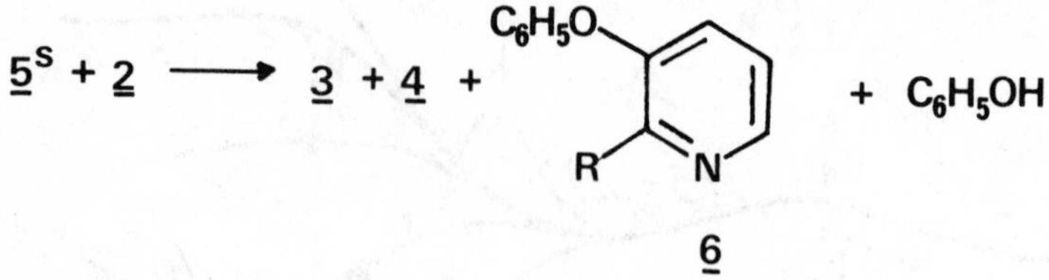

$$5^S + 2 \longrightarrow 3 + 4 + \text{(6)} + C_6H_5OH$$

**6**

a, R = CH$_3$

b, R = C$_6$H$_5$

When the *cis*-dimer, 4, is photosensitized in acetonitrile with 5, the *trans*-dimer 3, 6 and phenol are formed. At room temperature the *trans*-dimer is stable although both 3 and 4 quench the fluorescence of 5 at half diffusion rates. At room temperature, the photostationary ratio of *cis* to *trans* is > 99% *trans*-dimer. When 4 is photosensitized with 5 in acetophenone, the tetrahydropyran 7 is formed as well as the *trans*-dimer.

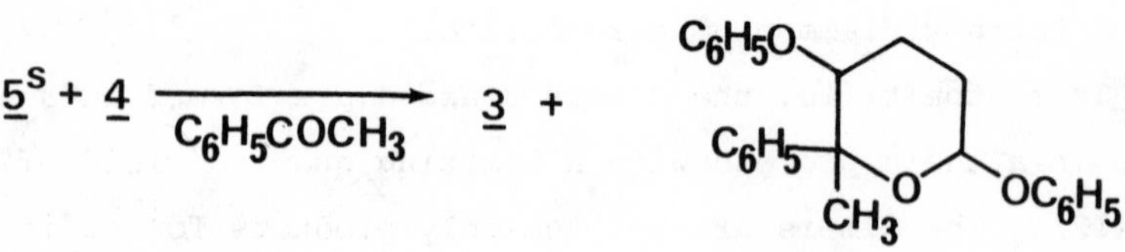

$$5^S + 4 \xrightarrow{\quad C_6H_5COCH_3 \quad} 3 + \text{(7)}$$

**7**

The mechanism that incorporates all of the experimental observations is

$$\underline{5} + \underline{2} \; \underset{R_{-d}}{\overset{R_d}{\rightleftharpoons}} \; (\underline{5}\cdots\underline{2})^s \; \overset{R_e}{\underset{R_r}{\diagup}} \; \begin{array}{l} \underline{5}+\underline{2}+h\nu \\[6pt] (\underline{5}^{\overset{.}{.}}, \underline{2}^{\overset{+}{.}})_s \rightleftharpoons \underline{5}^{\overset{.}{.}}_s + \underline{2}^{\overset{+}{.}}_s \end{array}$$

$$\underline{2}^{\overset{+}{.}} + \underline{2} \longrightarrow$$

$$\text{RCN} \quad \longrightarrow \quad \overset{C_6H_5O}{\underset{R}{\bigvee}} \text{OC}_6\text{H}_5 \quad \overset{+e^-}{\longrightarrow} \quad \overset{C_6H_5O}{\underset{R}{\bigvee}} \text{OC}_6\text{H}_5 \quad \overset{-H_2}{\longrightarrow} \quad \overset{C_6H_5O}{\underset{R}{\bigvee}} + C_6H_5OH$$

$$\underline{6}$$

$$C_6H_5COCH_3 \longrightarrow \quad \overset{C_6H_5O}{\underset{C_6H_5}{\bigvee}} \text{OC}_6\text{H}_5 \quad \overset{+e^-}{\longrightarrow} \quad \overset{C_6H_5O}{\underset{C_6H_5}{\bigvee}} \text{OC}_6\text{H}_5$$

$$\underline{7}$$

$$\underline{8}^{\overset{+}{.}}$$

$$\underline{3}^{\overset{+}{.}} \qquad \qquad \underline{4}^{\overset{+}{.}}$$

$$A^s \Big\Updownarrow +e^- \qquad A^s \Big\Updownarrow +e^-$$

$$\underline{3} \qquad \qquad \underline{4}$$

Thorough analysis of this reaction requires knowledge of all
relative energy levels. The first step of this reaction,
$\underline{5}^S + \underline{2} \longrightarrow \underline{5}^{\bar{\cdot}} + \underline{2}^{\bar{\cdot}}$, should be exothermic by *ca.* 10 kcal/mole
as determined from the relationship given by Rehm and
Weller (13), $\Delta G = \varepsilon_{\frac{1}{2}}(\underline{2}/\underline{2}^{+}) - \varepsilon_{\frac{1}{2}}(\underline{5}/\underline{5}^{-}) - \Delta E_{oo} - e^{2}/\varepsilon a$. The
relative levels for the reactions $\underline{2}^{+}_{\cdot} + \underline{2} \longrightarrow \underline{8}^{-} \longrightarrow \underline{4}^{+}_{\cdot}$ and
$\underline{3}^{+}_{\cdot}$ and/or $4^{+}_{\cdot} + \underline{2} \longrightarrow \underline{3}$ and/or $4 + 2^{+}_{\cdot}$ can be obtained by
determining the heats of formation and the oxidation potentials
of $\underline{2}$, $\underline{3}$, and $\underline{4}$, and the activation energy for the reaction
$\underline{3}^{+}_{\cdot} \longrightarrow \underline{8}^{+}_{\cdot}$. These values have been determined and used to
construct figure 4. There are a number of points that need to

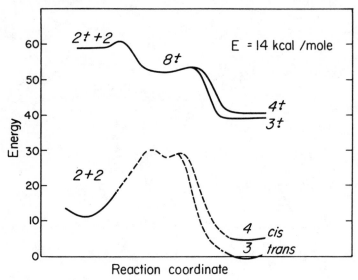

Fig. 4 *Potential energy diagram for the conversion of
phenyl vinyl ether into its dimer along both
the ground-state and the ion radical pathways.*

be made with reference to this diagram. The reaction, $\underline{2}^{+}_{\cdot} +$
$\underline{2} \longrightarrow \underline{8}^{+}_{\cdot}$ should be a rapid reaction since it is exothermic.
The rate constant for this reaction is found to be

$5 \times 10^8$ l mole$^{-1}$ sec$^{-1}$ (12). The radical cation $\underline{3}^+$ should open to $\underline{8}^+$ if the temperature is sufficiently high. At 60° C, $\underline{3}$, when sensitized with $\underline{5}$, does yield $\underline{7}$ in acetonitrile. The reaction $\underline{8}^+ \longrightarrow \underline{3}^+ + \underline{4}^+$ can be considered to be the energy gained in forming a one-electron bond. The relatively small amount of energy gained in this reaction is perhaps surprising. For the simplest one-electron bond, $H_2^+$, approximately 60% of the two-electron bond energy is present (14). The dissociation energy of unsubstituted and alkyl-substituted cyclobutanes, presumably the energy required to form the biradical RĊHCH$_2$CH$_2$ĊHR from the cyclobutane, is 60 - 63 kcal/mole (15, 16). Thus we might expect a value of ca. 30 kcal/mole for the reaction $\underline{3}^+ \longrightarrow \underline{8}^+$ rather than the observed value of ca. 15 kcal/mole.

However, the value of 60-63 kcal/mole is the value of a C-C bond reduced from ca. 80 kcal/mole for a normal C-C bond by the strain energy. Since all the strain energy remains in the one electron bond in cyclobutane we would expect a bond energy of $1/2 \times 80$ kcal/mole -20 kcal/mole $\cong$ 20 kcal/mole, as compared to the observed value of 15 kcal/mole.

Using the data compiled by Franklin et al. (17) on the heats of formation of ions and the ionization potentials of the radicals, along with the heats of formation of ethylene and cyclobutane (18) and the bond energy for the cylcobutane single bond (15, 16), we can construct the following diagram, figure 5 (19). Inspection shows that at the most, the stabilization of cyclobutane radical ion is ca. 9 kcal/mole. Therefore, we conclude that the one-electron bond energy for a cyclobutane single bond is only 15-25% of the two-electron bond energy and that this is characteristic of the cyclobutane ring.

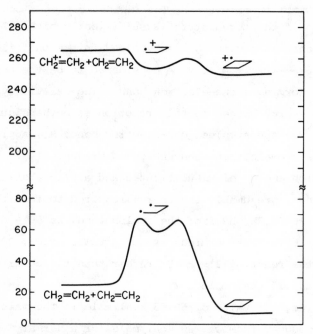

Fig. 5   *Potential energy diagram for the formation*
*of cyclobutane along both the ground-state*
*and the ion radical pathways.*

CONCLUSIONS

The photosensitized dimerization of phenyl vinyl ether
proceeds by an ionic mechanism in polar solvents. A good
yield of the dimers is obtained and the necessary require-
ments for a chain reaction are present. The first, $\underline{5}^{2 \cdot +}$
$\underline{2} \longrightarrow \underline{5}^{-} + \underline{2}^{+}$, and subsequent steps, $\underline{2}^{+ \cdot} + \underline{2} \longrightarrow \underline{8}^{+ \cdot} \longrightarrow \underline{3}^{+ \cdot} + \underline{4}^{+}$,
are all exothermic and the chain-propagating step, $\underline{4}^{+ \cdot} + \underline{2}$
$\longrightarrow 4 + \underline{2}^{+ \cdot}$, is only slightly endothermic (20).

# REFERENCES

1.  T.R. Evans, R.W. Wake and M.M. Sifain, *Tetrahedron Letters* 701 (1973).

2.  A. Ledwith, *Accounts Chem. Res.* 5, 133 (1972).

3.  J.A. Barltrop and D. Bradbury, *J. Amer. Chem. Soc.* 95, 5085 (1973).

4.  S. Tazuke, *J. Phys. Chem.* 74, 2390 (1970).

5.  N. Kornblum and F.W. Stuchal, *J. Amer. Chem. Soc.* 92, 1804 (1970).

6.  R.A. Neunteufel and D.R. Arnold, *ibid.* 95, 4080 (1973).

7.  S. Kuwata, Y. Shigemitsu and Y. Odaira, *J. Chem. Soc.*, D, 2 (1972); *J. Org. Chem.* 38, 3803 (1973).

8.  The fluorescence lifetimes for 9,10-dicyanoanthracene are 14.5 and 12.4 nsec (9) and the fluorescence quantum yields are 0.88 and 0.85 in acetonitrile and benzene, respectively. We have therefore assumed that the sensitizer lifetime is independent of solvent and have used the values for acetonitrile throughout this paper.

9.  D.R. Arnold, personal communication.

10. T.R. Evans, *J. Amer. Chem. Soc.* 93, 2081 (1971).

11. C. Reichardt, *Angew. Chem. Internat. Ed.* 4, 29 (1965).

12. S. Farid, S.E. Hartman and T.R. Evans *in* "The Exciplex" (M. Gordon, W.R. Ware, P. de Mayo and D.R. Arnold, eds.) Academic Press, New York, 1975.

13. D. Rehm and A. Weller, *Israel J. Chem.* 8, 259 (1970).

14. A.G. Gaydon, "Dissociation Energies and Spectra of Diatomic Molecules." 3rd Ed. p. 272. Chapman and Hall, Ltd., London, 1968.

15. M.R. Willcott, R.L. Cargill and A.B. Sears, *in* "Progress in Physical Organic Chemistry" (A. Streitwieser and R.W. Taft, eds.), Vol. 9, Chap. 2 Wiley-Interscience, New York, 1972.

16. P.J. Robinson and K.A. Holbrook, "Unimolecular Reactions." Chap. 7. Wiley-Interscience, New York, 1972.

17. J.L. Franklin, J.G. Dillard, H.M. Rosenstock, J.T. Herron, and K. Draxl, "Ionization Potentials, Appearance Potentials, and Heats of Formation of Gaseous Positive Ions." NSRDS-NBS, No. 26, National Bureau of Standards, Washington, 1969.

18. M. Kh. Karapet' yants and M.L. Karapet' yants, "Thermodynamic Constants of Inorganic and Organic Compounds," Ann Arbor-Humphrey Science Publishers, Ann Arbor, Michigan, 1970.

19. The uncertainty in the position of the open radical ion derives from the uncertainty in the free radical ionization potentials, which vary in an unpredictable manner, *i.e.*, ethyl radical, IP = 194 kcal/mole; propyl radical, IP = 187 kcal/mole; and butyl radical, IP = 199 kcal/ mole (17).

20. The oxidation potentials for 2, 3 and 4 are 1.75, 1.61 and 1.69 *V* respectively, in acetonitrile vs S.C.E. using a pyrolitic graphite electrode. Using a platinium electrode the oxidation potentials are 1.82, 1.81 and 1.81 Both electrodes show irreversible oxidation waves and the values are therefore suspect. The quenching constants for 2, 3 and 4 are 105, 105 and 112 l m$^{-1}$, indicating that the oxidation potentials are nearly identical for 2, 3 and 4.

INTRAMOLECULAR TRIPLET-STATE CHARGE-TRANSFER

INTERACTIONS IN AMINOKETONES

PETER J. WAGNER, DEAN A. ERSFELD AND B.J. SCHEVE

Chemistry Department, Michigan State University

East Lansing, Michigan  48824


INTRODUCTION

It now seems established that the three principal bimolecular  reactions between electronically-excited ketones and
other molecules are 1) hydrogen atom abstraction;  2) energy
transfer and 3) charge transfer.  The latter reaction is the
least well understood,  Both n-donors and π-donors quench
excited singlets and triplets of ketones, leading to different kinds of products with widely varying quantum efficiencies.  Olefins (1) and dienes (2) produce oxetanes in modest
to low quantum yields.  Amines (3) and sulfides (4) yield
photoreduction products, the former often in high quantum
yields, the latter with very low efficiency.  Benzene and
simple substituted benzenes quench triplet ketones but form
products in very low yields (5).

Quantitative kinetic analysis of CT quenching of excited
ketones reveals that full electron transfer is not involved
in the quenching step.  One line describes the relationship

between log $k_q$ and the ionization potential of quencher for amines, sulfides, thiols and olefins (6,7). The apparent activation energy for quenching rises only about 20% as fast as does the quenchers' IP. Moreover, rates of quenching are not significantly enhanced by polar solvents. It is probably safe to call the initial complex formed between excited ketone and these quenchers a CT complex. It may well be an exciplex, where most or all of the stabilization comes from electron transfer interactions. Unfortunately there is no evidence for exciplex emission or for dissociation back to excited ketone in any of the systems so far studied. The possibility remains that these quenching interactions do not lead to a real exciplex but rather to various high energy ground-state species with significant charge separation.

We have sought to obtain more information about these excited-state CT quenching processes by studying various bifunctional molecules in which the interaction can be intra-molecular. In such systems there are only limited variations possible in two important aspects of excited-state inter-actions: the distance between donor and acceptor and the mutual orientation of their orbitals. The systems which we have studied involve several phenyl alkyl ketones in which a tertiary amine functionality is incorporated in the alkyl group.

Tertiary amines quench the n,π* triplets of ketones such as benzophenone and acetophenone at rates which approach those of diffusional control (3,8). About half of these quenching interactions promote radiationless decay to the ground state of reactants; the other half yield radicals (3). The overall quenching rate constant is much too large for direct hydrogen atom abstraction, so the simplest mechanistic

$$R_2C=O^* + R_3N: \longrightarrow R_2\dot{C}-O^- \ \overset{\cdot\,+}{N}R_3 \longrightarrow R_2\dot{C}-OH + R_2\ddot{N}CHR'$$

interpretation postulates the formation of a CT complex which partitions its further chemistry between decay to the ground state and transfer of a hydrogen α to nitrogen to the carbonyl oxygen. The charge separation in the complex would certainly promote such a transfer. It is not known whether the complex dissociates to a pair of radical ions before proceeding to stable ground-state molecules. Since photoreduction quantum yields are high in low dielectric solvents such as benzene, it seems likely that the CT complex itself can rearrange to a pair of radicals. Hence the structure of the complex must be such that a hydrogen atom can approach to within bonding distance of the carbonyl oxygen.

We have previously studied some α-, γ- and δ-dialkylamino alkyl phenyl ketones (8). All of these clearly undergo considerable internal CT quenching in competition with Norrish Type II photoelimination. The behavior of the α-aminoketone is unusual in that most of the Type II reaction appears to be from the singlet. Kinetic analysis of the δ-aminoketone was quite straightforward, with intramolecular CT quenching (95%) dominating γ-hydrogen abstraction, which occurs at a rate consistent with those observed in other δ-substituted ketones. The γ-aminoketone also undergoes only Type II cleavage and cyclization. Again γ-hydrogen abstraction and CT quenching were postulated to be competitive. However, the hydrogens γ to the carbonyl are also α to the nitrogen, so that the excited CT complex could conceivably go on to a normal Type II biradical as in the bimolecular photoreduction scheme. In aprotic solvents, the Type II quantum yield is so low that little, if any, of the reaction could be proceeding *via* prior complexation. In methanol, however, the Type II quantum yield is so high that we speculated that the CT complex may well be partially rearranging to biradical. Unfortunately, it was impossible to differentiate quantitatively between direct and two-step γ-hydrogen transfer.

We have now studied two additional γ-dialkylamino ketones, γ-dimethylamino-2-butyronaphthone, 1, and 4-benzoyl-4,N-dimethylpiperidine, 2. The former possesses a π,π* lowest triplet which alters hydrogen abstraction tendencies; the latter keeps the nitrogen lone pair at fixed distances from the carbonyl.

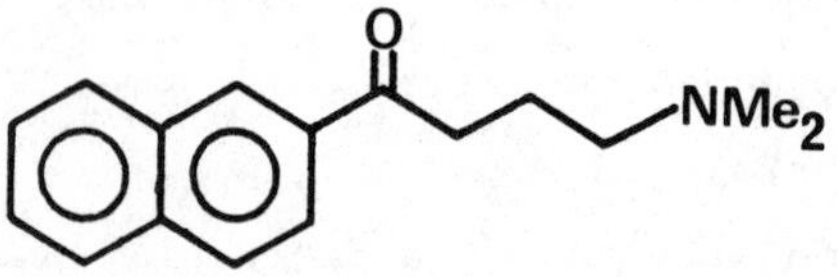

1 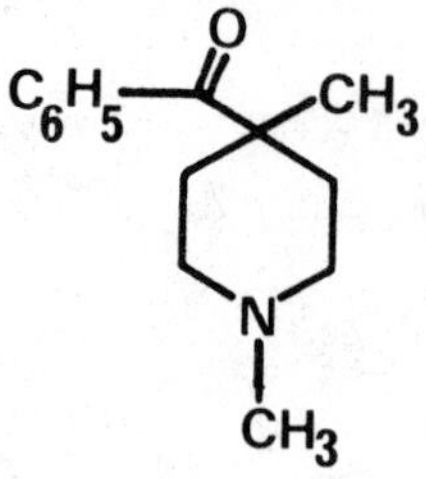            2

EXPERIMENTAL

*PREPARATION OF KETONES*

Ketone 1 was prepared by a standard Grignard addition of 2-naphthylmagnesium bromide to γ-dimethylaminobutyronitrile, followed by acid hydrolysis of the imine salt and distillation under reduced pressure. Compound 2 was prepared by a lengthy route which involved catalytic reduction of the methyl iodide salt of isonicotinamide, dehydration of the amide to the 4-cyanopiperidine, α-alkylation of the nitrile with phenylsodium and methyl iodide, and finally addition of phenyl Grignard to the nitrile. The structures of both ketones were confirmed by ir, uv, nmr and mass spectra. None of these spectra indicate any interaction between the carbonyl groups and the nitrogen.

*IDENTIFICATION OF PHOTOPRODUCTS*

The 2-acetonaphthone from 1 and the benzaldehyde from 2 were identified merely by their vpc retention times on several columns. The cyclic alcohol obtained from irradiation of 2 in the presence of naphthalene was recrystallized; its mass spectrum indicates that it is an isomer of 3. An hydroxyl group is prominent in its ir spectrum. Its nmr spectrum is consistent with the assigned, expected structure.

*QUANTITATIVE STUDIES*

All quantum yields, both relative and absolute, were measured by parallel irradiation of samples and valerophenone actinometers in a rotating "merry-go-round". An alkaline potassium chromate solution or Corning filters isolated either the 313 nm or 365 nm region, respectively, of a 450 W mercury arc. Samples were prepared in freshly purified solvents and were placed in 13 × 100 mm tubes which were then degassed and sealed prior to irradiation. Analyses of ketone disappearance, of product appearance and of pentadiene isomerization were performed by vapor phase chromatography, with alkanes as internal standards for the measurement of concentrations.

*$^{13}$C NMR SPECTRA*

These were obtained on a Brucker HFX-10 spectrometer at 22.6 MHz with the usual Fourier transform signal enhancement methods in conjunction with broad band proton decoupling.

Freon 21 served both as the solvent and as an internal lock.

RESULTS AND DISCUSSION

*COMPOUND 1*

Irradiation of degassed benzene or acetonitrile solutions 0.05 $M$ in $\underline{1}$ results in inefficient ($\Phi$ = 0.008), completely unquenchable (by neat 1,3-pentadiene) Type II elimination (9) to 2-acetonaphthone. The ability of $\underline{1}$ to sensitize the *cis-trans* isomerization of 1,3-pentadiene (10) indicates that $\underline{1}$ forms a long-lived triplet in high yield ($\Phi_{isc}$ = 0.80). In methanol, however, $\underline{1}$ undergoes Type II elimination with moderate efficiency, $\Phi$ = 0.11, 0.106 of which is readily quenchable by low concentrations of diene. Stern-Volmer analysis indicates a $k_q\tau$ value of 800 $M^{-1}$ and thus a $1/\tau$ value $\geq 4 \times 10^6$ sec$^{-1}$. This is the kind of value expected from the known rates of bimolecular quenching of triplet 2-acetonaphthone by triethylamine (11).

Since $\underline{1}$ possesses a $\pi,\pi^*$ lowest triplet which is known to abstract hydrogen atoms sluggishly if at all (12), it is unlikely that $\gamma$-hydrogen abstraction by the triplet carbonyl would compete with internal CT quenching. The small, unquenchable quantum yields for Type II elimination may represent either some singlet reaction competing with intersystem crossing or some reaction by the $n,\pi^*$ triplet in competition with internal conversion down to the lowest triplet. Since triplet $\underline{1}$ must undergo exclusive intramolecular CT quenching, the total lack of Type II reaction from the $\pi,\pi^*$ triplet demands that the CT complex be unable to proceed on to a biradical.

The CT complex probably involves close approach of the nitrogen lone pair to the carbonyl. The cyclic structure keeps the γ-hydrogen away from the oxygen. Moreover, the complex cannot dissociate into a freely rotating zwitterionic biradical, even in acetonitrile. Thus the ability of methanol to promote biradical formation probably involves a specific interaction with the complex such as a protonation of the somewhat negative carbonyl oxygen.

It is difficult to imagine any complex other than an exciplex which combines the requirements of close approach between nitrogen and carbonyl and only partial electron transfer.

*COMPOUND 2*

Irradition of degassed benzene solutions 0.04 *M* in 2 produces benzaldehyde and the bicyclic alcohol, 3, just as expected from the reported behavior of l-methyl-l-benzoyl-cyclohexane, 4 (13).

The benzaldehyde is the product of α-(Type I) cleavage; its formation is readily quenched by naphthalene and conjugated dienes, $k_q \tau = 260 \ M^{-1}$. Its yield is maximized by inclusion of the radical scavenger dodecanethiol in the solution. Not only is bicyclic alcohol formation not easily quenchable by typical triplet quenchers ($k_q \tau = 0.2 \ M^{-1}$), but modest diene concentrations (as well as the thiol) enhance its yield, such

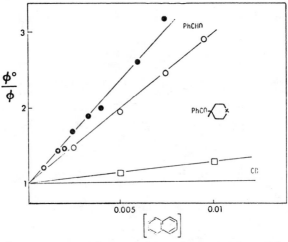

Fig. 1   *Quenching of benzaldehyde formation from*
*2 (x = NMe) ● and from 4 (x = CH₂) O;   of*
*bicyclobutanol formation from 4, □ .*

that at 0.5 *M* diene the conversion of 2 to 3 is nearly quantitative.   Only at quencher concentrations above 1 *M* does
noticeable quenching occur.   The α-cleavage reaction apparently both competes with cyclization and produces radicals which
react with 2 and 3.

As Lewis concluded for 4 (13), we also conclude that the
two conformers of 2 form distinct, noninterconverting triplets.   With the benzoyl group equatorial, 2e can undergo
only α-cleavage;   with the benzoyl group axial, 2a undergoes
very rapid γ-hydrogen abstraction followed by cyclization of
the resulting biradical.   Such hydrogen abstraction would be
maximally facilitated when the nitrogen lone pair is parallel
to the C-H bond being attacked.

Low temperature $^{13}$C nmr analysis of 2 indicates that the
ring flipping stops below -20° C.   The nitrogen inversion

366

does not begin to cease until $-85^\circ$ C. Comparison of the $-50^\circ$ chemical shifts of the ring carbons and the $\alpha$-methyl with their room temperature chemical shifts indicates that a 3:1 ratio of 2e:2a prevails at room temperature; a similar conclusion has been obtained for 4 (13).

The rate of triplet decay of 2e is slightly slower than that of the analogous 4e. The decay of 4e presumably corresponds almost totally to $\alpha$-cleavage, the rate of which should be the same in triplet 2e. Consequently we conclude that 2e undergoes negligible internal charge-transfer

quenching ($k_{ct} < 10^7$ sec$^{-1}$). Likewise the fact that 2 and 4
undergo Type II cyclization in comparable quantum yields
might suggest that the extremely rapid rate of decay of triplet 2a is due primarily to hydrogen abstraction and again
that CT quenching does not compete appreciably.

A closer look at the conformational possibilities of 2
reveals that the situation is considerably more complex.
There are four main conformers of ground-state 2 which can,
since conformational changes are slower than excited-state
reactions, yield four independent triplets. Nitrogen inversion allows the lone pair to be both axial and equatorial in
piperidines. Apparently there is no great difference between
the lone pair and a methyl group in conformational preference (14). Consequently we must consider that there are
roughly equal amounts of 2aa and 2ae, 2a with the lone pair
axial and equatorial, respectively. A like situation would
hold for 2ea and 2ee.

The Type I quantum yield of 2 is only half that of 4;
likewise the maximum Type II cyclization quantum yield of 2
is only one third that of 4. Approximately half the excited
states of 2 are unaccounted for. 2 sensitizes the *cis-trans*
isomerization of 1,3-pentadiene, as shown in figure 2. As
expected for a mixture of triplets with different lifetimes,
the usual reciprocal plot is not linear. The upper portion
corresponds to the long-lived triplet previously assigned
conformation 2e. Extrapolation to the intercept indicates
that this triplet is formed in only 35% yield, while total
triplet yield is near 100%. Since there could not be such
distinct triplet lifetimes if two triplet conformers were in
partial equilibrium and since 2ee and 2ea account for some
75% of the ground-state molecules, we conclude that only one
of them undergoes relatively slow α-cleavage while the other

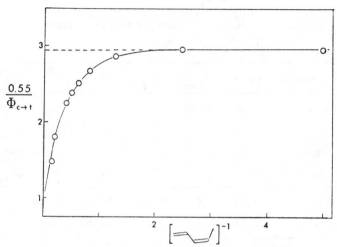

Fig. 2   Cis-to-trans isomerization of
1,3-pentadiene sensitized by 2.

must undergo some other faster, nonproductive decay.

The initial slope of the plot in figure 2 indicates a
$k_q\tau$ value of 0.2 $M^{-1}$ for some triplet species.  This probably
corresponds to 2aa, since 2ae could not possibly undergo such
rapid γ-hydrogen abstraction.  It is possible that a third
triplet with a lifetime of about 1 nsec is involved in the
sensitization plot, but the data are not sufficiently precise
to warrant such analysis.

Table 1 summarizes what we believe may be the kinetic
behavior of the various conformationally distinct triplets
of 2.  2ea undergoes only α-cleavage; 2aa primarily γ-hydrogen abstraction.  2ee and 2ae both undergo an unusually rapid
decay process which must involve the nitrogen.  Separate
experiments indicated that bimolecular quenching of one
aminoketone triplet by a ground-state aminoketone is too slow
to affect the measured triplet lifetimes.  Hence we speculate

TABLE 1

Ph — [structure] — X  *Excited-state kinetics*

| X | Conformer | GS % | $\Phi_{isc}$ | $\Phi_{\underline{1}}$ | $\Phi_{\underline{2}}$ | $10^8/\tau$, sec$^{-1}$ |
|---|---|---|---|---|---|---|
| NR | ea | 0.38 | 0.35 | 0.16 | 0 | 0.2 $(k_\alpha)$ |
| NR | aa | 0.12 | | 0 | 0.06 | 200 $(k_{\gamma-H})$ |
| NR | ae | 0.12 | | 0 | 0 | >10 $(k_{ct})$ |
| NR | ee | 0.38 | | 0 | 0 | >10 $(k_{ct})$ |
| $CH_2$ | e | 0.73 | | 0.30 | 0 | 0.25 $(k_\alpha)$ |
| $CH_2$ | a | 0.27 | | 0 | 0.18 | 2.0 $(k_{\gamma-H})$ |

that a through bond CT quenching can occur in $\underline{2}$ only when the
nitrogen lone pair is equatorial. This suggestion is somewhat akin to the well-known ground-state reactions of 4-
substituted piperidines which involve through-bond interactions (15).

CONCLUSIONS

This work provides further evidence for the required close approach between donor and excited ketone in CT quenching processes. This closeness prevails also in the resulting CT complex, which is probably an exciplex which reacts chemically too fast to allow phosphorescence or reformation of free triplet ketone. In compounds such as 2, in which the nitrogen lone pair is held 4-6 Å from the carbonyl, there is at least one conformation (ea?) in which CT quenching in negligible. This greatly reduced rate, compared to that in acyclic aminoketones, probably is due as much to poor relative orientation of orbitals as to distance between donor and acceptor. The quantum efficiencies with which 2 reacts and sensitizes suggests that something like through-bond CT quenching may occur. Other cyclic aminoketones are now being prepared to further test this notion.

## REFERENCES

1.  D.R. Arnold *in* "Advances in Photochemistry." (W.A. Noyes, Jr., G.S. Hammond and J.N. Pitts, Jr., eds.), Vol. 6, p. 301. Interscience, New York, 1968.

2.  J.A. Barltrop and H.A.J. Carless, *J. Amer. Chem. Soc.* *93*, 4794 (1971); N.C. Yang, M.H. Hui and S.A. Bellard, *ibid.* *93*, 4056 (1971); R.R. Hautala and N.J. Turro, *ibid.* *93*, 5595 (1971).

3.  S.G. Cohen, A. Parola and G.H. Parsons, Jr., *Chem. Rev.* *73*, 141 (1973).

4.  J. Guttenplan and S.G. Cohen, *J. Chem. Soc.* D 247 (1969).

5.  J. Saltiel, H.C. Curtis and B. Jones, *Mol. Photochem.* *2*, 331 (1970). D.I. Schuster and T.M. Weil, *J. Amer. Chem. Soc.* *95*, 4091 (1973).

6.  I.E. Kochevar and P.J. Wagner, *J. Amer. Chem. Soc.* *94*, 3859 (1972).

7.  J.B. Guttenplan and S.G. Cohen, *J. Amer. Chem. Soc.* *94*, 4040 (1972).

8.  P.J. Wagner, A.E. Kemppainen and T. Jellinek, *J. Amer. Chem. Soc.* *94*, 7512 (1972).

9.  P.J. Wagner, *Accounts Chem. Res.* *4*, 168 (1971).

10. A.A. Lamola and G.S. Hammond, *J. Chem. Phys.* *43*, 2129 (1965).

11. S.G. Cohen, G.A. Davis and W.D.K. Clark, *J. Amer. Chem. Soc.* *94*, 869 (1972).

12. G.S. Hammond and P.A. Leermakers, *J. Amer. Chem. Soc.* *84*, 207 (1962).

13. F.D. Lewis and R.W. Johnson, *J. Amer. Chem. Soc.* *94*, 8914 (1972).

14. H. Booth and J.H. Little, *Tetrahedron 23*, 291 (1967); J.D. Blackburne, A.R. Katritzky and Y. Takeuchi, *J. Amer. Chem. Soc.* *96*, 682 (1974); J.B. Lambert, D.S. Bailey, and B.F. Michel, *ibid.* *94*, 3812 (1972).

15. C.A. Grob, *Angew. Chem., Intern. Ed., Engl.* *8*, 535 (1969).